A Practical Approach to VLSI System on Chip (SoC) Design

Veena S. Chakravarthi

A Practical Approach
to VLSI System on Chip
(SoC) Design

A Comprehensive Guide

Second Edition

 Springer

Veena S. Chakravarthi ⓘ
Co-founder and Advisor
Sensesemi Technologies Private Limited
Bangalore, India

ISBN 978-3-031-18365-2   ISBN 978-3-031-18363-8   (eBook)
https://doi.org/10.1007/978-3-031-18363-8

This Springer imprint is published by the registered company Springer Nature Switzerland AG
The registered company address is: Gewerbestrasse 11, 6330 Cham, Switzerland

*An overview of the VLSI design
methodology and fundamental knowledge
required for a system designer to develop
complex System on Chip (SoC).*
                              −Veena S. Chakravarthi

Dedicated to VLSI designers

Foreword to the First Edition by Faraj Aalaei

It's an excellent time to be working in the semiconductor industry. Qualitatively, we are all familiar with Generation Z's constant appetite for digital consumption. That appetite is driving technical innovation starting in huge data centers and moving out to the growing sea of smartphones. Quantitatively, Gartner tells us that our industry is growing at a rate of 26% year over year. The semiconductor industry has never been more complex and it's going to keep getting more complicated. Every device needs to be smaller, more powerful, and more energy efficient than the previous generation.

There is no doubt our industry is shifting as waves of consolidation and innovation crash into new geographies and new markets, but the demand for intelligent, highly-integrated chip design keeps growing. This means that any aspiring hardware engineer – whether they want to work for a hungry, young startup or an established house of silicon – needs to become fully versed in the art of Very Large-Scale Integration (VLSI). There is no better teacher to learn from than Dr. Veena Chakravarthi.

I first met Veena in 2000 when she joined Centillium to play a key role in developing the high-performance System on Chip (SoC) solutions for Ethernet Passive Optical Networks (EPON). Those products helped us enable Asian service providers to deliver some of the first fiber to the home deployments in the world and threw fuel on the fire of data consumption. I've followed her career ever since, as she continues to add technical, professional and academic accolades to a stellar resume.

With thirty years of experience as a SoC architect and VLSI designer, Veena has distinguished herself as both an artist and an engineer. Her abilities to design large, complex electronics systems in silicon have created baseline, enabling technologies for a number of communications systems. Veena's depth of experience has allowed her to create a perfect primer for any engineer wanting to arm themselves with the necessary mindset to understand the chip design process and development cycle for SoCs. This practical approach contains straightforward applications of known techniques to create a structure which will help freshman engineers contribute effectively to the SoC design and development process.

I'm excited about the future of our industry and where SoC's can take us. They are at the heart of the advancements in medical, bio-tech, transportation, telecommunication and countless other industries that will change how we live. This book is a thoughtful guide for any aspiring chip designer, and I thank Veena for teaching the next generation of innovators, inventors and dreamers.

Faraj Aalaei is an Iranian-American immigrant, entrepreneur and 35-year veteran of the communications industry. He is the Founding Managing Partner of Candou Ventures. He was the CEO and Chairman of Aquantia Corp., a leader in the design, development, and marketing of advanced, high-speed communications ICs for Ethernet connectivity in the Data Center, Enterprise Infrastructure, Access, and Automotive markets. Prior to joining Aquantia, Faraj served as Chief Executive Officer and was one of the founders of Centillium Communications, a semiconductor solutions company. Before co-founding Centillium, he held a variety of positions with Fujitsu Network Communications and AT&T Bell Laboratories. Faraj has a Doctor of Engineering and a B.S. in Electrical Engineering Technology from Wentworth Institute of Technology, an M.S. in Electrical Engineering from the University of Massachusetts, and an M.B.A. from the University of New Hampshire. He holds three U.S. patents. An entrepreneur at heart, Faraj is a founding member and general partner at Monjeri Investment and Candou Ventures where he has investments in multiple start-ups in the communications, semiconductor, biotech, software, and e-commerce industries. As philanthropists, Faraj and his wife Susan Akbarpour are working to lessen inequality and to improve education inefficiencies around the world, serving as board members and supporters for multiple non-profit organizations. They are passionate about equality and well-being for children everywhere. Mr. Aalaei co-founded the Iranian American Political Action Committee (IAPAC) and serves on the board of Iranian American Contribution project (IACP) and Iranian Scholarship Foundation (ISF).

Founding General Partner, Candou Ventures   Faraj Aalaei
Palo Alto, CA, USA

Foreword to the First Edition by Ashok Soota

The semiconductor industry is undergoing a massive change with technologies like IoT, intelligent edge/cloud, mobility, automotive, 5G, AI and ML, creating in major opportunities. The expectations of 50 billion connected devices by 2025, and the massive amounts of data that will need to be processed on edge analytics as well as in the cloud, will result in sharper insights for better decision making.

With customers expecting continual improvements in applications, the question is whether the chip industry is moving fast enough to meet these expectations? A broad supply chain, equipment and materials innovations, and attracting the "best of the best" college graduates to fuel innovation are key.

This is an excellent time for young engineers to make the most of the opportunities and thereby fulfil their career aspirations, be it in corporate or Entrepreneurship. The book *Practical Approach to VLSI SoC Design* by Prof. Veena Chakravarthi is a good reference guide for new engineers and also a good refresher for seasoned practitioners of VLSI.

I have known Veena since early 2000, when she joined the core team of the technology business at Mindtree when she played a crucial part in developing successful in-house IPs like Bluetooth and WLAN core. Veena is a seasoned designer as well as an academician. Her experiences would be useful for both industry and academic needs and help engineers to take up path breaking design challenges.

Ashok Soota, Executive Chairman Happiest Minds, is widely recognized as one of the pioneering leaders of the Indian IT industry.

He was also Founding Chairman and MD of Mindtree, a company he led to a very successful IPO. Prior to Mindtree, he led Wipro's IT business for fifteen years. He also led the turnaround of Shriram Refrigeration into a highly profitable company after four straight years of losses.

Ashok has been the President of leading industry associations like Confederation of Indian Industry (CII), a member of the Prime Minister's Task Force for IT and on the Advisory Council for the World Intellectual Property Organization, Geneva. He is a Fellow of INAE and CSI and on the Board of Governors of Asian Institute Management (AIM), Philippines. He is a recipient of multiple IT Person of the year and Lifetime achievements awards.

Ashok's philanthropic contributions are channeled through Ashirvadam, a Trust he has created for environmental protection and help for the needy including vocational training, education and medical assistance.

Ashok is co-author of the National bestseller – *Entrepreneurship Simplified – From Idea to IPO.*

Executive Chairman, Happiest Minds   Ashok Soota,
Bengaluru, Karnataka, India

Foreword to the First Edition
by Walden C. Rhines

VLSI design of "Systems on a Chip", or SoC's, has suddenly taken a change in direction. Traditional computer architectures can no longer solve the computing problems of tomorrow. New, innovative approaches to SoC design will use non-Von Neuman architectural approaches with embedded neural networks to make problems like pattern recognition solvable in real time. Suddenly, the world of venture capital funded fabless semiconductor companies has exploded, as these companies propose innovative SoC's to solve "domain-specific" problems like vision, sound or smell related pattern recognition. Being able to do a few specific types of operations extremely well now becomes much more important than doing a wide variety of things very well. Beginning in the second half of 2017, the amount of venture capital money invested in fabless semiconductor and IP startups has accelerated, reaching an all-time record in 2018.

Books like *A Practical Approach to (VLSI) SoC Design* provide guidance for aspiring designers and academics who wish to join this parade of innovation. Rarely do opportunities like this emerge in the semiconductor industry. But this is a time of new ideas where the ability to translate algorithmic innovation to silicon can drive quantum steps forward in machine learning capability. The first wave of semiconductor technology was driven by physical component innovation. This wave will be driven by system innovation, combining unique software with clever hardware architectures. It will be an exciting revolution in computing.

Walden C. Rhines is President and CEO at Carnami. Prior to this, he was the CEO Emeritus of Mentor, a Siemens business, focusing on external communications and customer relations. He was previously CEO of Mentor Graphics for 25 years and Chairman of the Board for 17 years. During his tenure at Mentor, revenue nearly quadrupled and market value of the company increased 10X.

Prior to joining Mentor Graphics, Dr. Rhines was Executive Vice President, Semiconductor Group, responsible for TI's worldwide semiconductor business. During his 21 years at TI, he was President of the Data Systems Group and held numerous other semiconductor executive management positions.

Dr. Rhines has served on the boards of Cirrus Logic, QORVO, TriQuint Semiconductor, Global Logic and as Chairman of the Electronic Design Automation Consortium (five two-year terms) and is currently a director. He is also a board member of the Semiconductor Research Corporation and First Growth Children & Family Charities. He is a Lifetime Fellow of the IEEE and has served on the Board of Trustees of Lewis and Clark College, the National Advisory Board of the University of Michigan and Industrial Committees advising Stanford University and the University of Florida.

Dr. Rhines holds a Bachelor of Science degree in engineering from the University of Michigan, a Master of Science and PhD in materials science and engineering from Stanford University, a Master of Business Administration from Southern Methodist University and Honorary Doctor of Technology degrees from the University of Florida and Nottingham Trent University.

President and CEO at Carnami   Walden C. Rhines,
Dallas, TX, USA

Preface to the Second Edition

It has been four years since the first edition of this book was presented to readers with the intention of sharing practical methods of SoC design. The popular reception of this book, the advancement in the design techniques for the ever-growing complexity of SoCs, and the increased relevance demanded the new edition of the book. Edition 1 provides end-to-end SoC design methodology in much practical way as practiced in the semiconductor industry. This new edition

1. Provides technical corrections, updates, and clarifications in all eleven chapters of the original book and adds summaries of new developments and annotated bibliographical references at the end of each chapter;
2. Adds summaries of new developments with references at the end of each chapter;
3. Elucidates subtle issues that readers and reviewers have found perplexing, objectionable, or in need of elaboration.

Teachers who have taught from the first edition should find the revised edition more lucid while those who have waited for scouts to carve the path in VLSI careers will find the road paved and tested.

My main audience remain the students – students of electronics, electrical, and computer science who wonder why designing SoCs is a big deal; students who wonder how to convert the system requirements to working chips; students who are amazed about the advancements happening in electronics but not sure of how they are part of this wonderful journey. It continues to be an important reference book for researchers and career starters in VLSI design teams. I hope each of these groups will find the updated content presented in this book to be both inspirational and instrumental in tackling new challenges in their respective fields of SoC design. I thank all those who have gone through my earlier book and provided valuable feedback, which inspired me to author this edition.

Thank You.

Co-founder and Advisor   Veena S. Chakravarthi
Sensesemi Technologies
Bangalore, India

Praise for the First Edition

I am a system architect who works for a semi company and your book has helped me to have the big picture about SoC design. As you mentioned in the preface, this book is very close to industry practice and will help the engineer to understand the whole picture. A person like me who works as a system architect can benefit from this book to make the design flow become more complete. The evil is always in the detail, your book helps me cover it all.

Thanks
T J, System Architect

The book *A Practical Approach to VLSI System on Chip (SoC) Design – A Comprehensive Guide* by Dr. Veena S. Chakravarthi, truly lives up to its name. It is a practical guide to newcomers to the area of VLSI Design and comprehensive enough to be even used by industry veterans. I work in the VLSI design industry and teach design. I have seen this book being used effectively in both places. This very neatly organized book is easy to access for any topic and the examples make it very effective to quickly understand not only the concepts, but also quickly apply them.

The reissue of this popular book in its Second Edition with updated material should be an indispensable companion to practicing VLSI designers and students interested in pursuing a career in VLSI design.

Kumar M N
Chief Strategy Officer, Lead SoC Technologies

The book covers all necessary details of design and best practices to achieve power, area, and performance for career starters in VLSI. The basic design modules, scripts and the design flow explained in the book provide practical insight to the design process from RTL to Tape out. I am happy to be part of designing some of these examples for Veena. I would recommend this book to all the VLSI designers.

Vaibhav Rajapurohit
Senior silicon design engineer, Google

The book *A Practical Approach to VLSI System on Chip (SoC) Design – A Comprehensive Guide* by Dr. Veena S. Chakravarthi, provides more the than obvious SoC design method practiced. It prepares the reader for the changes necessary to meet ever-increasing needs of semiconductor chip designs. In the new era of intelligent computing and the large number of emerging applications like automotive, medical, IoT for SoCs, this book provides relevant ways to design them. We refer to this book quite often during our training activities.

Shivananda R Koteshwar
Group Director, Synopsys

Professionally I have known Veena for over 15 Years. I was closely associated with her during writing of the book *A Practical Approach to VLSI System on Chip (SoC) Design – A Comprehensive Guide*. I see it as an honest effort to bring design flow practiced in the industry into this book. I strongly recommend this to all VLSI designers.

Dinesh Annayya,
Director in Engineering - NEXT Edge Compute Engineering, Intel India.

This book by an author with decades of experience is sure to provide much-needed guidance to aspiring career starters, practicing engineers, faculty and students in VLSI. The very fact that this book is getting downloaded in large numbers is a testimony to its relevance. I hope the next edition of this book will cover all the recent advancements in the field of semiconductors which will further meet the expectations of people working in this exciting field.

Dr. K S Sridhar
Registrar, PES University

Preface to the First Edition

Having worked in semiconductor design industry for over two decades, it was my strong desire to pass on the knowledge of system on chip design to the next generation. Therefore, I conceived the idea of writing a book on **A Practical Approach to VLSI System on Chip (SoC) Design**.

The book intends to present a comprehensive overview of the design methodology, environment and requisite skills that are required for design and development of the System on Chip (SoC).

It ensures that engineers are aware and are able to contribute effectively to design companies from day one up to the development of complex SoC designs.

While this book is targeted at electrical and electronic engineers who aspire to be VLSI designers, it's also a valuable reference guide for professional designers who are part of development teams in VLSI design centers - the ones behind complex system-on-chip solutions.

The book aims to give readers a comprehensive idea of what one has to do as a VLSI designer. It expands on the arsenal of skills they need to be equipped with, the responsibilities of the job, and the challenges that they should anticipate. This information is based on my experiences in the semi-conductor industry and academics over the past twenty-five years.

Typically, electronic engineers aspire to become VLSI designers either during or after their undergraduate or graduate studies. Unfortunately for them, they usually don't possess the requisite skills and design techniques to circumnavigate the challenges they'll face in the industry. Meanwhile, young VLSI designers in the industry struggle to see the big picture of the design process. It's not practical for one person to work in all areas of the VLSI design and development process. This book is my attempt to provide answers to both groups, so that they can plan, understand, and equip themselves with the necessary skill sets. The design case relevance in every chapter and the design examples in chapter 11 help the readers realistically visualize problems and solutions encountered during VLSI system design.

The target audience for this book is engineering students who are pursuing a degree in electrical, electronics and communication and allied branches like bio-medical, biotechnology, instrumentation, telecommunication, etc. Also, engineers in the early stages of their career in the semiconductor industry can refer to the book for a complete understanding of the chip design process.

Though the books covers the complete spectrum of the topics relevant to System on Chip (SoC) using VLSI technology, it is good to have a fundamental understanding of logic design as it is a prerequisite to follow the contents of the book.

Though India is seen as a silicon country with Bangalore as a silicon city with many fabless design centers in VLSI, it is facing an acute shortage of employable VLSI design engineers as a large number of fresh engineers graduating from universities are not readily deployable for design jobs.

Statistics show that there is a demand for over 3000 design engineers per annum and that it will soon grow to over 30000 per annum in the coming years. Engineering schools currently cater to only 50% of the annual demand. Globally, the scenario is not too different.

In the current scenario of chip and resource shortage, VLSI design engineers have a promising and bright future ahead and can expect a challenging and rewarding career. According to Gartner's recent market research, the semiconductor industry is one of the fastest growing industries, at 26% annually, globally. And so are VLSI design jobs. Skilled VLSI people are always in demand, catering to the most challenging system on Chip designs, new versions of EDA tools addressing heterogeneous complex system integrations, Fabrication Technology correlations etc. Countries like Egypt need around 10,000 skilled VLSI designers.Globally, the semiconductor industry is one of the fastest growing industries at 26% annually. And so are VLSI design jobs. Skilled VLSI persons are always in demand, catering to the most challenging system on chip designs, new versions EDA tools addressing heterogeneous complex system integrations, Fabrication Technology correlations etc. Countries like Egypt need around 10,000 Skilled VLSI designers.

The design productivity gap - a shortage of skilled manpower that can convert transistors (that fabrication technology offers), to useful ones, is real. Hence there is a need to develop skill-sets to suit the semiconductor jobs and bridging this gap.

It would not have been possible to realize this project without the support of many of my friends, colleagues, and family. First, I wish to thank my father Mr. R S Chakravarthi, a noted journalist and Rajyotsava awardee from Karnataka, India, whose literary genes were responsible for harbouring my desire to write a book. My heartfelt thanks to my loving family, especially to Dr. K S Sridhar, my husband, K S Abhinandan, K S Anirudh, my sons; and Shradha Narayanan, my daughter-in-law for their support. I am indebted to my ex-colleague, Dr. M S Suresh, Scientist, ISRO, who patiently read each of my chapters and offered line-by-line reviews.

I wish to thank my ex-colleagues Mr. Sathish Burli for describing the software development flow and Dr. K S R C Murthy for sharing information on packaging with me. I thank my ex-colleague and dear friend Mr. Dinesh for identifying IOT-SoC reference design which is available in www.opencores.org for the case study. My steadfast team comprising of Vaibhav Rajapurohit, and my dear students Amruthashree, Aditya, tried out all the design examples and ensured that they are working and ready for the reference. Thanks to them.

I am also grateful to the semiconductor industry for having embraced me so warmly. And I'm mighty thankful to Mr. Faraj Aalaei, Founding General Partner, Candou Ventures, Mr. Ashok Soota, Executive chairman, Happiest Minds and Mr. Walden C Rhines, President and CEO at Carnami for taking time out of their busy schedules to write the foreword for this book.

I thank all the organizations and institutions I have worked with for contributing directly or indirectly to the naming of this book.

Last but not the least, I thank my superpower, who gives me the motivation and constant energy to take up projects beyond my capability and make them happen.

I will be very happy if users find each chapter useful and try out design examples and reference design and subsequently make VLSI their career choice. I am curious about your feedback and criticisms. I'm sure it'll go a long way in bettering this book.

Thank You

Veena S. Chakravarthi is a Bangalore-based technologist, system-on-chip architect and educator. Over a career spanning three and half decades, she has spawned of several VLSI design & incubation centres and managed several high-performance tech-teams at ITI Limited and across various MNCs like Mindtree consulting private limited, Centillium India private limited. Transwitch India private limited, Ikanos communications private limited, Periera ventures, Asarva chips and technologies, Sankhya labs, Prodigy technovations private limited and Synopsys, India She holds a PhD from Bangalore University and an MPT certification from IIM Bangalore.

About the Book

Why read this book?

The main goal of bringing out this book is to present the end-to-end design flow of a system on a chip (SoC) in the comprehensive manner possible. This book is targeted to students of undergraduate and graduate courses in computer science and electrical and electronics engineering, and new hires in VLSI design teams. It can also be an important reference for professional designers who are part of development teams in VLSI design. It aims to give the readers a complete perspective of what one must do as a VLSI designer, the skillset required, the job content, and the challenges faced in chip design. The information is based on the author's personal experience in the semiconductor industry, whose academic career spans over three decades.

What Problems Does It Solve?

Typically, engineers during their undergraduate and graduate courses aspire to become VLSI designers but lack knowledge of the necessary skillset for designing complex SoCs. Paradoxically, VLSI designers in the industry will not have a big picture of the SoC design process end-to-end, as chip design is a very complex and specialized process. This book attempts to provide answers to both groups so they can plan, understand, and equip themselves with the necessary skill sets.

Who is the audience?

Students of electrical engineering, electronics and communications engineering, and computer science, as well as students of allied fields such as biomedicine, biotechnology, instrumentation, and telecommunication, are the intended audience for this book. Also, engineers in the early stage of their careers in the semiconductor industry can refer to the book for a complete understanding of the chip design process to get the complete process of the design and development cycle of System on Chip.

What are the prerequisites to reading this book?

Though the book covers the a complete spectrum of topics relevant to System on Chip (SoC) using VLSI technology, it is good to have a fundamental understanding of digital design, working knowledge of Linux platforms, and scripting languages as a prerequisite.

Why become a VLSI designer?

Though India is seen as a silicon country, with Bangalore as a silicon city with many fabless design centers using VLSI technology, it is facing an acute shortage of employable VLSI design engineers. Reports show a demand for more than 3,000 design engineers per year, which will soon rise up 30,000 per year in the coming years. With the shortage of chips and countries signing strategic plans for funding semiconductor operations, it is an excellent opportunity to be part of the growing industry. In this scenario of chip shortage, a VLSI design engineer has promising and bright career prospects, with a challenging and technically satisfying career.

Globally, the semiconductor market is estimated to hit $1 trillion by 2030, and the Indian semiconductor market is expected to grow at a CAGR of 16% to reach $64 billion in 2026, according to a report by IESA. Skilled VLSI personnel will be required to cater to the most challenging SoC designs using the latest EDA tools, for complex system integrations and to address the challenges of submicron process technologies. This means the design productivity gap exists. There is an urgent need to develop the skilled resources to bridge this gap.

Contents

Abbreviations

ADC	Analog to Digital Converter
AHB	Advanced High-performance Bus
AMP	Asymmetric Multiprocessing
API	Application Program Interface
ASIC	Application Specification Integrated Circuit
ASCII	American Standard Code for Information Interchange
ATE	Automatic Test Equipment
ATPG	Automatic Test Pattern Generation
ATSE	Advanced Television Systems Committee
BCL	Base Class Library
BGA	Ball Grid Array
Bi-CMOS	Bipolar Complementary Metal Oxide Semiconductor
BIST	Built in Self-Test
BS	Boundary Scan
BFM	Bus Functional Model
CIF	Caltech Intermediate Format
CMOS	Complementary Metal Oxide Semiconductor
CSP	Chip Scale Packaging
CTS	Clock Tree Synthesis
CVD	Chemical Vapor Deposition
DAC	Digital to Analog converter
DDR	Double Data Rate
DEF	Design Exchange Format
DFT	Design for Testability
DMAC	Direct memory Access Controller
DRC	Design Rule Check
DRM	Design Rule Management
DRV	Design Rule Violation
DUT	Design Under Test
ECO	Electronics Change Order
EDA	Electronic Design Automation

EM	Electro Migration
ERC	Electric Rule Check
ESD	Electro Static Discharge
EU	Effective Utilization
EVM	Electronics Validation Module
FCS	Frame Check Sequence
FBGA	Fine Pitch Ball Grid Array
FET	Field Effect Transistor
FPGA	Field Programmable Gate Array
FPU	Floating Point Unit
FSM	Finite State Machine
FIFO	First In First Out
FTP	File Transfer Protocol
GALS	Globally Asynchronous and Locally Synchronous
GDS II stream format	Graphic database system II stream format is industry standard format in which the IC design layout
GSLA	Globally Synchronous and Locally Asynchronous
HDL	Hardware Description Language
HFN	High Fanout Nets
HLD	High-Level Design Document
IC	Integrated Circuit
IEEE-SA	Institute of Electrical and Electronics Engineers Standards Association
I2C	Inter-integrated Circuit
ICG	Integrated Clock Gate
I2R	Input to Register
I2O	Input to Output
IO	Input-Output
IP Cores	Intellectual Property Cores
ISP	In-System Programming
ITU-T	International Telecommunication Union-Telecommunication
JTAG	Joint Test Action Group
LAN	Local Area Network
LBIST	Logic Built in Self-Test
LC	Inductance-Capacitance
LEC	Logic Equivalence Check
LEF	Library Exchange Format
LFSR	Linear Feedback Shift Register
LIB	Liberty File Format
LINT	Tool that analyze programming and flag errors based on set of rules defined.
LVS	Layout Versus Schematic
MBIST	Memory Built in Self-Test
MCM	Multi-chip Module
MIL	Military

MIPS	Million Instructions per Second
MRD	Market Requirement Document
MISG	Multiple Input Sequence Generator
MEMs	Micro Electro Mechanical Systems
MoCA	Multimedia over Coax Alliance
MSV	Multiple Supply Voltage
MSSV	Multi Supply Single Voltage
NAS	Network Attached Storage
NRE	Non Recurring Engineering
OCV	On Chip Variation
OS	Operating System
OSCG	On-SoC Clock Generation
PCB	Printed Circuit Board
PGA	Pin Grid Array
P&R	Place and Route
PR Boundary	Place and Route Boundary
PRD	Product Requirement Document
PRPG	Pseudo Random Pattern Generator
PTAM	Power Aware Test Access Mechanism
PLL	Phase Locked Loop
PMBIST	Programmable Memory Built in Self-Test
PVD	Physical Vapor Deposition
PVT	Process-Voltage-Temperature
RC	Resistance-Capacitance
RTL	Register Transfer Level
ROI	Return on Investment
R2R	Register to Register
R2O	Register to Output
SAN	Storage Area Network
SEM	Scanning Electron Microscope
SDC	Synthesis Design Constraint
PDP	Preferred Data Path
SDF	Standard Delay Format or Synopsys Delay Format
SI	Signal Integrity
SIP	System in Package
SLEC	Sequential Logic Equivalence Check
SMD	Surface Mount Device
SSN	Simultaneous Switching Noise
SPEF	Standard Parasitic Exchange Format
SPI	Serial Peripheral Interface
SPICE	Simulation Program with Integrated Circuit Emphasis
SMP	Symmetric Multi-Processing
SoC	System on Chip
SRAM	Static Random-Access Memory
STA	Static Timing Analysis

STUMP	Self-Test Using MISR and Parallel SRPG
TLF	Timing Liberty Format
TPI	Test Program Interface
TSMC	Taiwan Semiconductor Manufacturing Company
QFP	Quad Flat Package
UART	Universal Asynchronous Receiver/Transmitter
USB	Universal Serial Bus
UV	Ultra Violet
UVM	Universal Verification Methodology
VHDL	VLSI Hardware Description Language
VIP	Verification Intellectual Property
VLSI	Very Large Scale Integration
WIFI	Wireless Fidelity
WSP	Wafer Scale Packaging
WNS	Worst Negative Slack
Fab-less	Companies which do all services except the wafer and chip fabrication process

Chapter 1
Introduction

1.1 Introduction to CMOS VLSI

VLSI stands for very large-scale integration. Complex systems on silicon chips are developed using VLSI technology. The most dominant VLSI processing technology is CMOS VLSI. The system uses many millions of transistors on a chip. Transistors are basic device elements that are used to develop many complex systems. The system is realized on a semiconductor chip and is hence called system on chip (SoC). The advantages of CMOS VLSI SoCs are small size, low power, and high speed. Smartphones, electronic gadgets, and infotainment products use SoCs. CMOS is the dominant VLSI technology. CMOS technology for decades obeyed Moore's law. Moore's law states that "the number of transistors in a chip doubles every 18 months." This has proved correct ever since 1965. But this posed, and continues to pose, innumerable challenges to the designers.

1.2 Application Areas of SoC

System on chip (SoC) has become an indispensable part of many products in every domain. Applications in the communications, data storage, and high-tech computing domains traditionally use SoCs. Advanced SoCs of today have even penetrated into medical, automotive, infotainment, security, and defence applications. SoCs are used in almost all applications that need signal, data, computing and communication. Some of the visibly noted applications are the following:

- All IOT applications in healthcare, automotive, home automation, and industry automation.
- Large data centers and data farms.
- Smartphones and mobiles.

© The Author(s), under exclusive license to Springer Nature Switzerland AG 2022
V. S. Chakravarthi, *A Practical Approach to VLSI System on Chip (SoC) Design*,
https://doi.org/10.1007/978-3-031-18363-8_1

- Medical devices.
- Satellite communications and space technologies.
- Agriculture automation technologies.
- Multimedia infotainment.

1.3   Trends in VLSI

System on chip industry has developed rapidly over the last few decades. Systems have become complex, dense, heterogeneous, large, robust, and highly powerful with low-power consumption. The design methodologies have matured and are most complex. The future trends in growth of VLSI technology are discussed under the following heads:

- System on Chip complexity.
- Performance Power and Area (PPA) Goals of SoC.
- Size of the SoC Dies.
- Design Methodology.

1.4   System on Chip Complexity

Since the time transistors were invented in the past six decades, the physical dimension of the transistor has been constantly shrinking. This has resulted in packing more and more transistors on a silicon wafer, integrating more and more functionalities into the circuits. This phenomenon is called *scaling*. And this is still continuing. But it is predicted that in the next couple of years, the scaling of transistor's dimensions will reach a point where it will be so expensive that it will become commercially not viable to scale down further. This is partially true. The cost of fabrication has grown exorbitantly, making design and development more critical to make it correct by design. Technological advancements have continued to scale down the features further. Recently, the semiconductor foundries have announced 3 nm technology for commercial SoC development. And researchers claim the success in developing technologies to realize device features of 1 nm [1]. This shows that even today, there are researchers who believe that scaling down the feature size is possible using technologies beyond CMOS technologies and alternative materials to the most commonly used ones. This appears to be more promising than existing methods for developing highly integrated systems on a chip (SoC). Scaling down to an atomic size (0.1 to 0.5 nm) is currently thought to be impossible. But the fact of the matter is that it is scaling that has driven the tremendous growth in the SoC industry in computation and communication applications. Processor processing power has increased many folds, changing the way we sense, process, store, display, and communicate information of any magnitude. Over the past couple of decades, the complexity of chips has increased from simple time critical circuits to multiprocessor-multicore systems. Today's electronic products need very few off the chip components, apart from the system on chip (SoC) unlike the products of yester year. The trend in integrating more and more circuits to form SoC was the result of advancement in VLSI allied technologies. Advancement in CMOS VLSI fabrication processes have enabled the development of the most complex system on chips possible. This was accelerated by the feature enhancements in EDA tools with intelligent algorithms. Cerebras' wafer-scale engine (WFE) chip with 2.6 trillion transistors—that's 2,600,000,000,000 with 850,000 cores on TSMC 7 nm process technology—was successfully fabricated in 2019. The largest general-purpose processor with 114 billion transistors and largest graphic processor unit (GPU) with 80 billion transistors are commercially available. Other types of ICs, such as field-programmable gate arrays (FPGAs), have the largest transistor count of around 50 billion transistors with 9 billion logic cells. This demonstrates the complexity of the SoCs of today (Fig. 1.1).

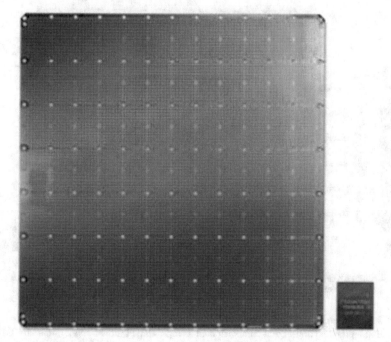

Fig. 1.1   Cerebras' wafer-scale engine. (Courtesy: Cerebras)

1.5   Integration Trend from Circuit to System on Chip

VLSI designs in 1970s were small time critical circuits and required to work with standard general-purpose processors to realize integration on printed circuit boards (PCBs) with many other devices. During the earlier days of design, the time critical circuits were schematically drawn in the EDA tool and were interconnected with other modules. The advancement of CMOS technologies, packing more and more transistors in a small area, and the invention of the automated synthesis tool (converts the design representation using hardware description language into schematic), made it possible to define large complex designs for complete systems. The scaling phenomenon, an advancement in process technology and the improved design methodologies enhance the compatibility of non-digital circuit fabrication to CMOS fabrication, enabling the integration of non-digital components. The integration can be on the package containing ICs (technology called SIP) or on to a chip as system on chip (SoC). Non-digital components include RF, analog, and sensor devices. Figure 1.2 depicts the International Technology Roadmap for Semiconductors (ITRS) trend of integrating digital and non-digital components into a single chip.

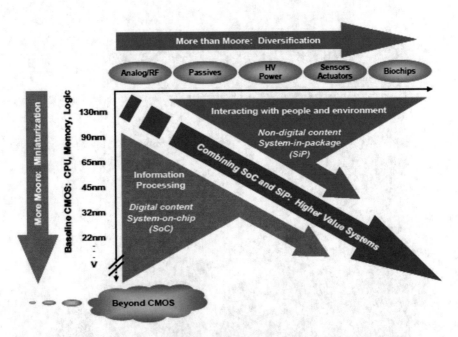

Fig. 1.2 The International Technology Roadmap for Semiconductors depicts the ITRS trend of integrating digital and non-digital components in a single chip as a dual trend: miniaturisation of digital functions ("More Moore") and functional diversification ("More-than-Moore"). (Source: ITRS white paper)

The International Technology Roadmap for Semiconductors (ITRS) has empha-sized that scaling in CMOS technology and its associated benefits in terms of per-formance will continue. This direction for further progress is labelled "More Moore." The second trend is integrating non-digital functionalities to contribute to the miniaturization of electronic systems, although they do not necessarily scale at the same rate as digital functionality. This trend is named "More-than-Moore" (MtM).

Advances in EDA tools made it possible to realize complete systems on chip by means of automation and analysis capability. SoC modelled with its behavioural description in hardware description language (HDL) is converted to schematics by synthesizing and the design process is called physical design was able to generate the design database (this database is in GDS II format, and the process of submitting the database to the fab is called tape out) which is used directly in the fabrication process of chip. In the present day, VLSI designs are all system on chip designs of great complexity. The complexity of the SoC chips ranges from simple microcon-troller systems to large networks on chips utilizing hundreds of millions of transis-tors. Figure 1.3 shows the evolution from a simple circuit on chip to a system on chip (SoC).

Today's SoCs, for example, smartphone SoCs like Qualcomm's snapdragon series, contain embedded processors like an ARMv8 processor, general-purpose processor, a DSP, RF transceiver, WLAN 802.11 ac cores, embedded memories, cache, and analog interfaces. Each of the SoC's functional cores WLAN 802.11 ac core and RF transceiver, is controlled by one or more embedded processors of vari-ous complexities. Another example is Intel's i-series chips, which contain multiple processor cores that can function independently and have fast interface cores com-plying with interface standards like PCI Express, USB, and on-chip memories.

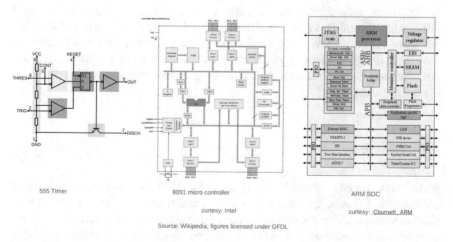

Fig. 1.3 Complexity trend in ICs

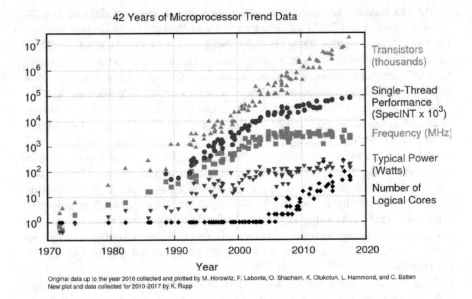

Fig. 1.4 Complexity trends in computation system on chip

1.6   Speed of Operation

Another trend observed over the last six decades is the phenomenal increase in the speed of system. Figure 1.4 shows the trends in speed, power, transistor density, and the number of logic cores. High-speed systems on chips (SoCs) developed by leading semiconductor companies claim to operate at a frequency of 2.5 to 3 GHZ. Also, few of the system on chips support the data transfer rate of 100 GB/s. All these trends offered many challenges to the designers, and this resulted in changes in design methodology over the years. These challenges are responsible for devising new design methods and modelling at the high abstraction levels of system hierarchy.

1.7   Die Size

As the transistor size shrank, more and more transistors were packed in smaller area on a silicon die; thus, the transistor density (number of transistors per unit area of silicon) increased. This resulted in realizing more and more functions in a small area of the die and enabled the realization of complete coordinated functions of the system to be designed on the die. To cope up with Moore's law, die size increased by 14% every two years (Source: Intel), thus enabling the realisation of a complete system on chip (SoC). Thus, began the era of miniaturization, which resulted in

Fig. 1.5 Generations of computers. (Source: IBM)

generations of computers ranging from main frames to personal computers of high performance. Figure 1.5 shows three generations of computers made by system on chips (SoCs).

Today's high-performance smartphones which can be carried in pockets are the results of this miniaturization and integration of non-digital functional blocks in the VLSI technology.

1.8 Design Methodology

To complement the advancements in VLSI technologies over the past six decades, the design methodology has evolved over the years. This was made possible by the availability of large computing resources and the development of design automation tools. These tools can be considered as linchpin technologies, which are major enablers for complex SoC design. Examples are synthesis tools, simulators, STA

Fig. 1.6 EDA tools complementing the technological advancements

tool, and physical design tools. Figure 1.6 depicts the EDA tools complementing the technological growth by using computerized automatic methods in place of hand designs. Further, the design productivity gap instigated the virtual design core developments and made reuse an inevitable choice in the large designs of today. During this time, the design entry methods changed from simple schematic entry to the interconnection of many functional design cores of processors and peripherals (called intellectual property cores—IP cores). The intellectual property core is a functional block that can be bought on licencing or royalty terms. Once bought, it can be reused multiple times. The number of intellectual property (IP) cores being integrated is close to a hundred and more in present-day systems. Enabler to this advancement is also the high computation capability of workstations and systems, which enabled large design databases and verification by simulations.

The choice of design methodology for a SoC depends on conflicting factors: performance (in terms of speed or power consumption), cost, and volume. Major design options are custom design, standard cell-based design, and array-based design. A complex SoC design may employ any or all of these options.

1.9   SoC Design and Development

With the advancement of technology, the design and development environment of SoC is constantly upgraded with newer advanced skill sets, intelligent tools with advanced algorithms, standard design guidelines resulting in more predictable chip performance, modeling and hardware description languages, high-capacity development systems operating at high frequency of the order of tens of GHz, large memories of the order of multiples of terabytes and large processing power. This demanded human resources with a new skill set.

1.10   Skill Set Required

The skillset required in the VLSI designer changed from circuit fundamentals to ability in realizing the the functionalities by logic definitions and modeling using hardware description languages as the design complexity and methodology changed over the past couple of decades with advent of intelligent EDA tools. The major hardware description languages used to describe the hardware functions are Verilog, SystemVerilog and VHDL. This should be supported by knowledge of the tool usage to get the desired functionality by guiding the tools with proper input of the design description files and constraints. It is important for the designer to have fundamental knowledge of chip design and design flow. Scripting languages like TCL-tk, Perl will come handy in automating the simulation, synthesis, and STA scripts which are to be run iteratively and when reports and logs generated by design tools are to be analyzed. Most importantly, imagining the hardware and then coding for it helps in hardware realization and debug. Flexibility to work in any department of design like logic design, synthesis, timing analysis, physical design, and FPGA validation, makes a designer the most desirable.

1.11   EDA Environment

As the design complexity evolved from time critical circuitry to system on chip, the algorithm-based tools for synthesis and timing analysis, and physical design tools like placement and routing, got developed and matured to the extent that the tools were able to write out design databases for the most advanced fabrication technology interface which mask is making equipment. The design database is used to make masks based on advanced optical and electron beam lithography. Simultaneously, the verification methodologies such as Universal Verification Methodology (UVM) supported by cycle-based and event-based simulators and advanced hardware description language like SystemVerilog help to succeed in the development of complex SoCs in first attempt. Validation results of these chips show great correlation to design simulations. Major EDA tools used during SoC design are simulators, synthesis tools, static timing analyzer, P&R tools, parasitic parameter extractors, and formal verification tools such as equivalence checkers, design and electrical rule checkers, etc. Looking at the complexity of SoCs, there are tools powered by machine learning and artificial intelligence (AI) that help the designer to take right design decisions at every level of design flow. Some tools are already being used in placement stage of SoC design. For advanced technology nodes, there are also tools which take in the design database and generate mask data with necessary process parameter corrections which help in getting processing done first time right. FPGA-based development, which was initially viewed as competition to VLSI development began to be viewed as complementing the VLSI design process for first-time SoC success.

1.12   Challenges in All

Trends and advancement discussed in previous sections shows that they require constant upgradation of the skills and techniques to adapt to the fast-changing fabrication technology by scaling and design methodology in terms of tool usage and system modeling. In addition, electronic products are, as they are characterized by obsolescence, driving shorter development cycles and shorter time to market. This drives VLSI designer to be on their toes to be smarter, efficient, and knowledgeable about the advancements in tools to be able to contribute to the development of system on chip. Technically, the more and more integration of the functional blocks and its realization by fabrication using CMOS and CMOS-compatible technologies results in a lot of on-chip variations, resulting in huge challenges to achieve large yield and SoC performance. Debugging bad SoCs is extremely challenging. Power management is another major challenge of today's SoC. It is essential to have innovative power management designs to curtail power consumption, good quality power regulation, and conversion efficiency. Packaging technologies like SIP pose the challenge of good quality integration and power management and can become the alternative to SoC.

Reference

1. 'Taiwan's TSMC claims breakthrough on 1nm chips' News reports

Chapter 2
System on Chip (SoC) Design

2.1 Part 1

2.1.1 System on Chip (SoC)

System on chip (SoC) is defined as the functional block that has most of the functionality of an electronic system. Very few of the system functionalities, such as batteries, displays, and keypads are not realizable on chip. CMOS and CMOS-compatible technologies are primarily used to realize system on chips (SoCs). MEMs-based sensor technology, Bi-CMOS technology, and memory technology are some of the CMOS-compatible technologies.

Present-day SoCs are far more complex. They contain hundreds of processor cores, tens of DSP engines, and many interface and protocol IPs connected on the high-speed bus as a network on chip cores. Typical SoCs contain the most essential functional blocks of products. In addition, they have on-chip memories and test and diagnostic cores, which help to make them robust and reliable.

2.2 Constituents of SoC

A typical SoC consists of the following functional cores:

- Several general-purpose RISC processors.
- One or more DSP processors.
- On-chip embedded memory/memories.
- On-chip, protocol blocks.
- Controllers for external memory for increased memory.
- One or more standard interface controller cores like USB and PCIe cores.
- On-chip clock generation block or clock recovery and stabilization logic blocks.

© The Author(s), under exclusive license to Springer Nature Switzerland AG 2022
V. S. Chakravarthi, *A Practical Approach to VLSI System on Chip (SoC) Design*,
https://doi.org/10.1007/978-3-031-18363-8_2

- Power management and distribution networks.
- Analog cores.
- User interface blocks like keyboard and display controllers and communication cores and radio interfaces.
- Sensor and attenuator conditioners or controllers.

In addition, a SoC has embedded software with software loading for booting functionality with default configurations from factory setup. This is why, when you buy a product, it works right away without the need for any additional software installations or setup. Each of the constituents of SoC can be designed independently by different design and development methods. The design methods such as automated cell-based, full-custom design flows (analog, mixed signal blocks, phase-locked loop (PLL) circuits, pad circuits), MEMs design flows, and structured array-based design flow (embedded memory) can be adopted when designing them independently. They are then integrated as single chip or multiple dies stacked and packaged.

2.2.1   Processor Subsystem Cores

Most SoC with on-chip cores include single or multiple processors. A core is the smallest unit of processor capable of running instructions on its own and having the ability to interact with other functional blocks within the SoC. Processors are used for different on-chip control and data processing functions in the SoC and to configure and control peripheral devices. One such example is a Bluetooth transceiver in an Internet of Things (IOT) SoC with its own processor core to configure and function as per the Bluetooth protocol for its various functional modes. The multimedia SoC generally has an on-chip processor to process the media data/signal as required by an application.

The on-chip processor has embedded software for its operation. Multicore processors pose an interesting problem from a software point of view, the major challenge being sharing the processing load in executing the functionality among them and coordinating to achieve the overall and each core's individual performance in a SoC. Figure 2.1 shows an example of one of the architectures of multiprocessor cores in a SoC. Some of the most commonly used architectures are the following:

- Asymmetric multiprocessing (AMP): In this mode, the SoC architect partitions the processor SW functionality for each of the cores. This ensures that each of the processors has different programs residing in the SoC. Each core is independent in a way, and runs its own software, and has an exclusive memory space. Cores may run an operating system (OS) or direct code without an underlying OS. The software code that runs directly without the operating system (OS) is called bare metal. Each core will have its own set of interrupts and access-specific peripherals. Processors in a SoC communicate with each other through shared

Fig. 2.1 SMP-AMP
processor structures

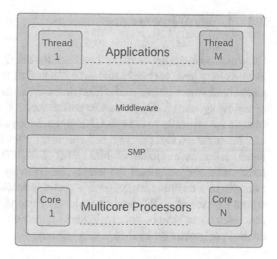

memory or interrupts. All these were planned while architecting the SoC at the beginning of the development. This includes decisions on memory sizes, types, and the number of interrupt lines and their types.

- Symmetric multiprocessing (SMP): In this mode, the operating system (OS) is allowed to decide the best core to run the job on. This also implies that all the processor cores are generic, and it cannot be determined which core is running a particular job—it can vary based on the real-time status of the cores. In SMP mode, the address space of the processors is shared, i.e., all the cores can access a common memory area because, based on the load conditions, any of the cores may be asked to do a specific job. Sharing of data is done via memory, which is controlled by the OS. SMP modes are typically used when the jobs are generic and the need is a computation resource. Most of the cloud SoCs are of this type.

Based on the application of the SoC, processors can be divided into the following categories: application processors and control processors.

2.3 Application-Specific Processors

These are typically high-performance computation engines, that run SoC-specific applications and control the interfaces in the SoC. They tend to run operating systems like embedded Linux, Android, etc. Most application processors are multicore. They are driven by clock frequencies ranging from a few hundred of MHz to several GHz. Application processors typically run in SMP mode. One such example is smartphone SoCs.

2.4   Control Processors

Control processors are used for functions that are tightly coupled with the hardware. They usually have very tight real-time constraints and need to respond back to associated hardware within specific time limits. Most control processors run a real-time operating system (RTOS) to ensure performance. Clock frequencies for control processors are typically in the sub-GHz range. Many control processors also have custom instructions, that are designed specifically for the identified functions. Each of these custom instructions combines a set of steps into one single instruction, which accelerates and optimizes the use of the software for its performance. One such example of a custom instruction is a cyclic redundancy check (CRC) computation, where a series of XOR steps could be combined into one instruction. The control processor cores have custom registers to improve performance. Most of the on-chip control processors typically run in AMP mode.

2.5   Digital Signal Processors

SoCs are designed for applications that require fast specific signal processing functions such as FFTs, encoding and decoding of bits, and interleaving and de-interleaving operations. Digital signal processing (DSP) cores offer specific instruction sets, that are suitable for this type of processing. This allows designers to embed DSP cores and do the signal processing functions in software rather than hardware. From a SoC point of view, DSPs can be considered as control processors. They typically have their own memory areas and communicate with other processors using shared memory or interrupts. DSP cores in SoC are typically programmed in bare-metal software.

2.6   Vector Processors

In many SoCs, there are very specific tasks which are too small to add a control processor or a DSP and, at the same time, are best not done in hardware for flexibility purposes. For example, consider an encryption algorithm, which may have to be changed, based on region the SoC is being sold in. In such a case, designers would like to have a small core, which they can load with the specific algorithm based on the region. This would keep the SoC generic. Region-specific adaptation could be done via software, rather than designing SoC variants or putting all the hardware into the SoC. Vector processors can be considered as mini-DSPs which are loaded and initialized on the go by one of the other processors in the SoC. They are always bare-metal codes. There are many commercially available SoCs.

2.6.1 Embedded Memory Core

Embedded memories are static RAMs, which are available as hard macros with wide range of configurations for a particular technology. The desired size and orga- nization can be configured using a memory compiler tool provided by the memory vendors. Memory compiler can generate all design relevant files with views of the selected memory model that are required for the design stages. It can write out an RTL model for simulation, timing models for timing analysis, and structural frame views for physical design steps. Apart from the size of the memory, different con- figurable parameters supported by the memory compiler are aspect ratio, number of sub-banks, and row and column address multiplexers depending on the desired (PPA) performance like access time, power consumption, and area. Memory com- piler can also include BIST circuitry and peripheral circuitry such as redundant bit addition and error correcting code (ECC). On-chip compiled memories designed in this manner are optimized for performance but are fixed for a chosen configuration. As a design guideline, small memories of sizes ranging from 1 to 10Kbytes are designed using flip-flops as register arrays. Medium-sized memories are typically SRAMs. A typical SRAM memory layout is shown in Fig. 2.2.

For larger memories, there are compiled memories available that are optimized for area, power, and speed of operation. These memories are of single transistor memories called "1 T memories", suitable for SoCs. These memories are denser and are used in memory-rich SoCs for data storage on the chip. Level 1 and Level 2 memories used as cache memories are typically SRAMs in processor SoCs.

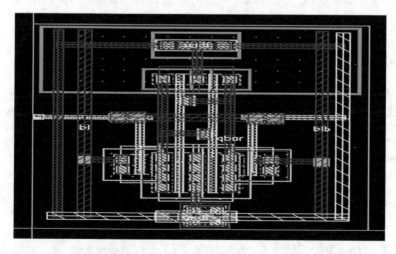

Fig. 2.2 SRAM memory cell layout

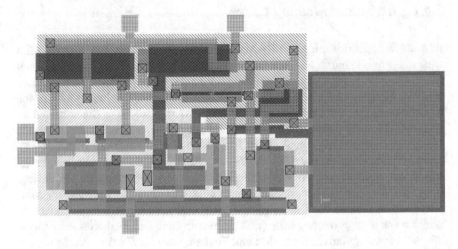

Fig. 2.3 Layout of OP-AMP

2.6.2   Analog Cores

Analog cores like OP-AMPs, transceivers, power amplifiers, serdes, phased-lock loop (PLL), and mixed signal blocks are typically found in most common SoCs. These are designed separately using mixed signal design procedures and are integrated as hard macros in the SoC during the physical design stage. The simulation models of the analog cores are interface timing models, which are verified in full SoC level simulations. As an example, the layout of OP-AMP is shown in Fig. 2.3.

2.6.3   Interface Cores

Another important constituent of the SoC is the various interface or communication blocks which provide connectivity of SoC to external peripherals or devices for greater onboard integration. Some of the examples of interface cores are USB, UART, SPI, DDR2, PCIe, AXI, and AHB master/slave controllers. These cores commonly have local processors and embedded software performing the protocol functions of these cores. These cores are either designed in house or third-party IP cores are available in hard or soft cores.

2.6.4   On-Chip Clock Generators, PLLs, and Sensors

In addition to the SoC constituents that are responsible for functional operations, there are some necessary on-chip functional blocks which are required for the SoC operation. These are:

- On-chip clock generators.
- PLLs.
- On-chip sensors.

2.7   Part 2

2.7.1   SoC Development Life Cycle

Product requirement is captured by the market research in the area of business for a company. Market research is the study to identify the customer's needs and provide technology solutions in the area of company's objective. A probable technology solution is detailed to identify the system requirements and feasibility of its development as a system on chip or SoC solution. For example, an agritech company may identify SoC solutions for drones to be applied for surveillance of agricultural lands, crops, and remote monitoring. However, the SoC development is considered only after a thorough understanding of the market demand and development feasibility. Once decided to develop a SoC for the wireless drone controller, a technical study is carried out to capture the functional requirements. A functional specification is then derived from the functional requirements. Typical functional specifications of such a SoC include transceiver specifications, range of control, regulatory compliance requirements, protocol standards applicable, power requirements, etc. These details are documented in the market requirements document (MRD) with a preliminary estimate of development and manufacturing costs. This is the first step in the product development cycle. From the MRD, the requirements for the product and system are derived and documented as product requirements document (PRD). PRD documents the application scenarios and identifies various functional cores that are required to be integrated in the proposed system. It defines electronics hardware system requirements, peripheral modules, user interfaces, casing, etc. The electronic system is then mapped to the targeted appropriate technology for development, and this is when the system on chip SoC is visualized. The system architect further studies the feasibility of the design within the engineering and cost constraints. This is an iterative process involving many reviews and cross-functional discussions between marketing and systems groups. Once accepted, PRD is passed to the engineering team for studying the feasibility of development. The system on chip is targeted at the target VLSI and related technologies. SoC is then partitioned to subsystems which are then identified, for in-house modules or off-the-shelf IP cores. A decision on suitable general-purpose processors is taken after technical feasibility, which sometimes involves technical experimentation and analysis. Further, functions that need special signal processing functions requiring dedicated digital signal processors or modules are identified which gives input for actual hardware-software partitioning of the system. All these are highlighted in the high-level design document (HLD) of the system. It is from here that the engineering design teams plan and start to design and develop the SoC.

2.8   SoC Design Requirements

HLD for a SoC lists explicit functional specifications, detailing the applicable technology standards, certification needs, and packaging requirements. The technology standards to be followed during SoC development for safety or interoperability are defined by professional bodies such as IEEE, ATSE, or ITU-T, among others. SoC performance factors (PPAs), like power, area, and speed, become the chip vendor's unique selling proposition. Hence, it is essential to identify innovative ways to design SoCs to meet the set design goals.

2.9   Design Strategy

SoC design approach depends on a number of factors, including type of SoC (digital, analog, or mixed signal), PPA (power performance and area), EDA tool flow, and if it is the first-generation or subsequent design versions. In most of the SoC designs, the functional cores that are of high value to the company are developed in-house, and the rest of the general-purpose IPs are bought from third-party design vendors and integrated. It is essential for every designer to be aware of the strategy to align his/her role in the design and development of SoC. The commercial viability of the SoC depends on design complexity, PPA needs, target process technology, and the volume planned to be produced. Design and development of high PPA SoCs is expensive. To be competitive in the market, it is required to minimize the cost of development. SoCs used in consumer applications are very cost sensitive and must be competitive. Achieving low cost and high PPAs for SoC are possible only if they are manufactured in large volumes. However, there are exceptions: SoC requirements in strategic areas like defense and space applications demand very high performance but are needed in small numbers. In such cases, the cost of the SoCs will be small fraction of cost of large systems. However, in all categories, the goal of designers is to minimize nonrecurring engineering (NRE) costs and aims to achieve a high PPA for SoCs. When SoC is in production, the cost per part will be a function of die size, targeted fabrication technology, packaging, testing, and validation where the economy of scale works. The larger the production volume, the lower the cost. All these require the right strategy for designing the SoC.

2.10   SoC Design Planning

SoC design starts with the development of chip architecture. Chip architecture contains descriptions of functional blocks and their interfaces, communication interface blocks, the clock and reset strategies, power up and booting procedures, data paths, control paths, and intellectual property (IP) cores. It is to be noted that complete clarity on all the details may not be available at this stage, but this initial document serves as a starting point. System architects develop this document by experimenting with and modelling some of the blocks if required. This also forms the basis for resource planning, defining tool flow, and design infrastructure planning. Chip architecture is used to decide on the number of designers, and verification engineers, number of design workstations, networking needs, and EDA tools required as planned. Design planning in new design houses starts with one or two modelling engineers and modeling/simulation tools, and the rest of the requirements are topped up eventually in due course of development. A chip architecture document is used to define the design flow and methodologies depending on the SoC complexity and performance requirements. Custom design flow is adopted when SoC performance needs are very high, and an automated cell-based design flow is adopted for digital designs. Automated cell-based design flow is also called the standard cell design. Analog blocks like high-speed data converters, clock generation circuits, PLLs, and high-performance serializer/de-serializer (serdes) are the candidates for custom design methods. Custom design methodologies are not suitable for large SoC designs. For such SoC designs, standard cell-based design flow is the right choice. In this approach, a library of pre-designed standard cells of a wide variety of logic gates over a wide range of drive strengths is used. The standard cell library contains all the standard logic cells and some of the cells such as adders, comparators, encoder-decoders, and clock buffers. Most of the EDA tools support this flow. The standard cell approach has become a de facto industry standard for large complex SoC design. Deciding the composition of the cell library has become a crucial activity at present while adopting the right design strategy.

2.11   System Modeling

In the HLD and chip architecture, system blocks are identified with few design assumptions in terms of processing time, algorithm choice, latencies, and clocking data path throughputs, which are validated by modelling the subsystem. A system reference model is developed and is used as a golden reference against which the actual design is verified. The most commonly used languages for modelling systems are high-level programming languages like C++, System C, System View, MATLAB, and Scilab. The system model reassures the correctness of partitions, interfaces, and algorithms to be used in the SoC design.

2.12   System Module Development Feasibility Study

Though system models validate the implementation possibility to an extent, hardware design constraints may restrict achieving the set SoC design goal. The feasibility of achieving the PPA goals for SoC is evaluated by alternate implementations on development platforms and emulation methods to select the right implementation techniques. Some of the decisions on parallel and serial functions like cyclic redundancy check (CRC), frame check sequence (FCS), and memory requirements are decided by these methods.

2.13 IP Design Decisions

SoC will have functional processes and tasks, that are executed on the general-purpose processors and processor subsystems. General-purpose processor cores are part of most of the SoCs. These cores are designed for high-performance and optimum area. Typically, there are companies which design the processor or processor subsystem cores exclusively for varied performance. They offer these cores in many forms, as IP cores for SoC integration. These are to be validated for performance and latencies as required by the integrated systems. Application-specific SoC designers buy processor cores and subsystem cores, DDR controllers, and standard protocol interface cores as IPs for integration. These are proven IPs for functionality, interoperability, and integration. The IP cores are bought on royalty or license terms by SoC designers. Availability, reuse, and portability of soft macro modules to any target technology enables SoC designs to be carried out and offered at a faster time to market. Processor cores, security engines, and interface IPs like USB, UART, SPI, and HDMI are examples of such readily available IP cores.

2.14 Verification IPs

Like design IP cores, verification intellectual property (VIP) cores are pre-modeled and verified soft cores which can be integrated to SoC verification environments. This helps to uncover the compatibility and misinterpretation of functionalities of the IP cores in SoC designs. Verification IPs are available on royalty or licence terms and can be reused in the verification of multiple SoC designs. VIPs are offered along with a set of standard test scenarios that help to verify the SoC designs faster. Examples of VIPs are SPI master/slave cores and Ethernet MAC cores. Because most verification IPs are not be synthesizable, they are used only for design verification.

2.15 Target Technology Decision

Once the processor subsystems and the IP cores and suitable packages are identified, the next step is to identify the target technology for SoC processing. This decision is primarily driven by the power budget for the chip, estimated die size, and availability of the identified IP cores like PLL, memories, and other essential cores in the target technology. It becomes a business decision if the IP cores are not proven in the targeted technology as it increases the SoC design time. This is an iterative decision-making process until a suitable target technology decision is made.

2.16 Development Plan

SoC architecture identifies all the required subsystems and IP cores, schedules are worked out, and depending on the time to market (as decided in the MRD), design tape out plan is made. Accordingly, a make or buy decision is made for IP cores, keeping in mind the early market entry advantage for the companies. However, the third-party IPs may require some design wrappers around them to integrate them into the system and validation to check the suitability of the integration. Apart from the standard cell library from the target technology vendors, SoC design generally requires some additional complex cells called macro/mega cells, which are available off the shelf even by the EDA tool vendors. Examples of macro cells are "fast multipliers" and memory arrays. Macro cells can be reused in many future designs, which can offset the initial design cost. The functional macros are of the hard macro or soft macro types. Hard macros are cores that are available as designs which can be integrated into the SoC at the physical design stage. The SoC designer cannot modify them in any way but can only connect them to the SoC internal blocks through the input-output signals. Major advantage of using hard macro is that the macro cell is optimized for PPAs, in terms of size, power dissipation, and speed. The hard macro has the disadvantage of not being portable to other process technologies. But generally, for parametrizable hard macro cells, the vendor provides a macro generator which can be used to generate the macro cell of the required parameters as configurations. For example, from a memory compiler, it is possible to generate a wide variety of memory arrays of different sizes. Soft macros are functional modules with predetermined functionality and are available as a synthesizable core. Soft macros can be ported to any process technology chosen for SoC design. Soft macros are integrated prior to the synthesis stage of the SoC design. This must go through the synthesis and physical design processes to meet the SoC design goals. They can be customised to suit the SoC integration. Example is a multiplier module.

2.17 EDA Tool Plan

EDA tools play a very important role in the SoC design process. EDA tools are used for functional and timing verification, design synthesis; DFT; physical synthesis; place and route; and design rule checking during SoC design. They can be broadly classified as design tools and verification tools. Major design tools are synthesis tools and place and route tools. Major verification tools are functional and timing and fault simulators; static timing analyzers; equivalence checkers; design rule checker; and electrical rule checkers. There are many tool vendors who offer end-to-end EDA tools with an easy-to-use graphical user interface for SoC design. Even then, it is necessary to customize the tool flow based on the SoC design complexity and type of design. Typically, SoC designs are carried out with EDA from one vendor for design implementation and another set of tools for verification and analysis. This is because of the complexity involved in SoC design and the belief that the algorithms used for VLSI design implementation are different from those used in design verification. This helps to find any mismatches in design during verification. Apart from the EDA tools, there are other support tools for design database management, debugging, and analysis tools. The minimum set of tools required to design a SoC are HDL simulators, LINT checkers, synthesis tools, static timing analyzers, sign-off tools, and sometimes hardware-software co-simulators, FPGA validation setup and modelling softwares. It is also required to have a design repository management tool with revision control and bug tracking capability. The custom design flow require extraction and modeling tools, circuit simulator layout editors, design rule checkers, and electrical rule checkers in the design environment.

2.18   Design Center Infrastructure

SoC design is a computation-intensive process requiring high-performance systems for design simulations, synthesis, and physical design processes. Depending on the design complexity, process run times vary from a few minutes to multiple days during the design cycle. This requires high-end server level workstations and design systems with the right operating systems on which the design processes are run. Most of the SoC designs are carried out on Linux-based high-performance computer systems. The SoC design process is also a team effort where many designers access the different sets of tools at different points of time in the design cycle. This requires a proper network of systems with the right access rights provided to the systems and tools to the designers. It is also important to have proper backup facilities and security of the IP database as it is of high value design process. A typical network setup for SoC design is shown in Fig. 2.4.

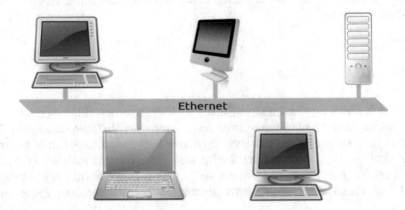

Fig. 2.4 Design infrastructure network topology

2.19   Computational Servers

Computational servers (servers that perform heavy duty computations) are the machines that run the simulations, logic, and physical synthesis of SoC designs. These machines have configurations which are geared for the needs of tools which actually run high-end design processes. A typical machine could have 8–16 cores running at 2 GHz or more and working with 64GB of memory (RAM) or more. It also required a large sized cache for holding temporary data during design transitions from input to output formats. The EDA processes also generate a large amount of data. The waveform output of a simulation could reach 100 GB or more.

2.20   Filers

A storage filer is a file server designed and configured for high-volume data storage, backup, and archiving. Storage filers are also known as network-attached storage (NAS) filers or storage area network (SAN) filers. They are useful when a lot of data has to be shared across multiple users across ethernet LANs.

The best storage filers are characterized by around-the-clock availability, scalability, expandability, and ease of management. They typically support multiple network protocols and have high storage capacity. Many of them support storage redundancy, high throughput, security features, and connectivity to a variety of backup device types and configurations.

2.21   Workstations

Workstations are high-performance systems with good graphics capabilities, large storage, and powerful multiple processors that are used by VLSI designers. As personal laptops come with these capabilities, designers use high-performance laptops for most of the design phases. Workstations are used for final layout editing for fixing design rule checks and other guideline violations during physical design verification. The major considerations for choosing the workstations are the EDA tool requirements and design complexity.

2.22   Backup Servers

A backup server is a type of server that enables the backup of data, files, applications, and/or databases on a specialized in-house or remote server. It combines hardware and software technologies that provide backup storage and retrieval services to connected computers, servers, or related devices. A backup server is generally implemented in an enterprise IT environment where computing systems across an organization are connected by a network to one or more backup servers. A backup server consists of a standard hardware server with substantial storage capacity, mostly with redundant storage drives, and a purpose-built backup server application. The backup schedule for each computer may be installed with a client utility application or configured within the host operating system (OS). At the scheduled time, the host connects with the backup server to initiate the data backup process. The backup may be retrieved or recovered in the event of data loss, data corruption, or a disaster. In the context of a hosting or cloud service provider, a backup server is remotely connected through the Internet on a Web interface or through vendor application programming interfaces (API).

2.23   Source Control Server

Important component in the design center infrastructure is source control server which helps to manage the revisions of the source code developed as the design database. It is also called revision control, or version control server. This is the main server which hosts the design database and its modifications, and the logs of changes made by design owners and their details like time of change, from which login details and over time of design. Changes to documents or source code are identified by the revision numbers or tags. The corresponding database with the tag can be retrieved if required at any point of time. This enables tracking of the changes in the database from the initial version or revision till the final version. This also helps in the release mechanism to transfer the database from one group to another in a multi-team environment consisting of a design team, verification team, synthesis team, and physical design team. These systems and the software support database tagging, merging, backing off the changes, etc., but the operation on the database will be recorded and hence provide traceability.

2.24   Firewalls

Firewalls are hardware or software systems which prevent unauthorized access to the repository server or source control server. It is very important to have the control mechanisms to access systems where SoC design databases are stored and processed.

2.25   Resource Planning

Good design is possible by the great designers. Designers with the right skill set and expertise can only succeed in designing high-quality SoCs first time. Design teams working on complex SoC designs require different skill sets depending on their roles. System architects must have complete system level knowledge, different algorithms and be able to interpret standard specifications. In addition, they must also understand clocking strategies, low-power consumption concepts, processor architectures, bus structures, memory organization, and their effect on the perfor-mance. They must have good understanding of modelling techniques and have a working knowledge of design and verification requirements. Front-end or logic designers must be good at the fundamentals of logic design, concepts of synthesiz-ability, HDL programming, timing concepts, and design flows be aware of the rel-evant EDA tools. For good SoC design, it is essential to have a good mix of designers, verification engineers, implementation engineers, tools experts, network support teams, and physical design teams. Also, in the design team, it is required to have expert designers in digital and analog circuits, with good protocol understanding depending on the SoC requirements.

2.26   SoC Design Flow

The SoC design flow involves multiple parallel design flows adopted for different design subsystems. Complete subsystems are integrated into one SoC design either at the logic, synthesis, or physical design stage. Finally, the design database is taped out as a single SoC design for fabrication after complete verification and sign-off. The following sections explain different design flows adopted for different types of designs, such as processor cores or subsystems, digital subsystem cores, analog/RF cores, or memory controllers.

2.26.1   SoC Chip High-Level Design Methodology

Since the last six decades, the design methodologies have evolved so much that the focus is shifted to system designs from circuit designs in VLSI technology. VLSI design flow has become a small part of the entire system design, and the approach to system development has become more of an integration of many of the functional blocks in complex systems. System design is a set of subsystem design flows exe-cuted in parallel and integrated at various stages. Major design and development flows are listed below:

- Digital SoC core development flow.
- Processor subsystem design flow.

- SoC physical design flow.
- Software development flow.
- EVM/SW development platform design flow.
- Product integration flow.

2.26.2   Digital SoC Core Development Flow

Digital SoC core development flow is standard ASIC design flow or standard cell design flow. The digital SoC core of the SoC is used to design a critical part of the system which is the main differentiator in an application. The automated standard cell-based design flow is shown in Fig. 2.5.

The functional specification is defined by the critical block around which the overall system on chip is planned. The standard cell design flow is used if this block is a fully digital logic core. This block is functionally partitioned into sub-blocks, and each sub-block is defined in detail. This is called a design document or *micro-architecture* design. This can be at the module/submodule or chip top level depending on the complexity. The design details of any submodule or module include the functional description, internal block diagram, interface signal description, their timing diagrams, and internal state machine details, if any. The design document specifies some special features and scenarios which are critical to the functionality that needs to be verified. This information is typically used as a requirement to create special functionality or capability in the test bench to verify the design scenarios. For example, in the design of a circular buffer of 1 K locations, when the data is continuously getting written and read out, it is not normal to get the buffer in full condition unless the read is stalled. This is the design corner in this context. It means that it is required to stop reading the buffer to see if the buffer is getting full and test if further data written is properly getting written to the start of the buffer as it is circular without losing the last data being written. Figure 2.6 illustrates the design corner condition of the circular buffer.

As per the microarchitecture, the module/block or chip core is behaviorally modelled using hardware description languages like SystemVerilog and VHDL. SystemVerilog has now become the de facto hardware description language for chip design. This is the RTL design stage. Here, the functional block is designed using register transfer language (RTL) using HDL. RTL design must comply with design guidelines to be able to accept it for further design processes like synthesis, which is a critical design step in the SoC design flow. The HDL modelled design is verified for the correctness of its functionality by simulations using the test bench using simulators. Simulators are software tools that support application of test vectors and capture the responses of the design under test and generate graphical signal waveforms for designer to analyze and decide for correct behavior. Simulators are of two types: cycle-based and event-based simulators. Most commonly, for digital design, cycle-based simulators are used. The RTL design is then synthesized with proper design constraints. Design constraints are the designer's requirements with clock

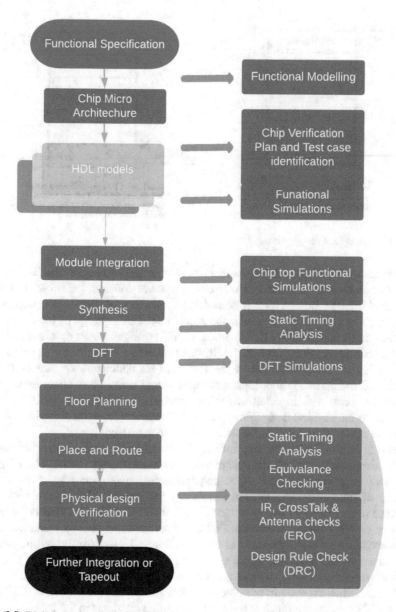

Fig. 2.5 Digital core standard cell design flow

details, input and output delays, and instructions to use specific logic synthesis instructions that are used by the synthesis tool. It contains tools to use standard logic cells in the technology cell library to interconnect them in a particular way to meet certain area, timing, and power goals of the design. Synthesis is the process which reads the HDL behavioral modules and converts it to gate level design abstractions

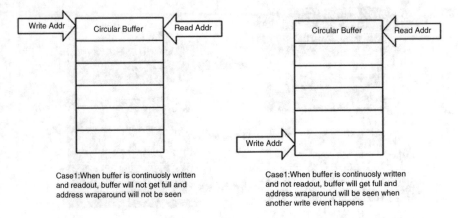

Fig. 2.6 Example of design corner

called design netlist. Netlist representation of a design is a set of standard gates/ cells/flip-flops interconnected to realize a particular function described in HDL model of the RTL design. This is done using synthesis tool. During the process of synthesis, the D flip-flops inferred in the design netlist are replaced by the scan flops for the design for testability (DFT) process. DFT is the process of ensuring that the module failures resulting from the fabrication process are traceable and identifiable. The design is further modified by the DFT tool for additional test structures for embedded memories, D flip-flops, and input-output pads. More about these processes are dealt with in detail in further chapters. A final design netlist is then released to the physical design flow, which is normally referred to as backend flow. Physical design flow converts the design netlist to the design layout.

The floor plan is the first step for the physical design, which is the placement of the submodules and all the design elements in the design netlist. The IO pad placements, power requirements, on-chip memories, macros, and the interconnect ability of the submodules within placement and routing (PR) boundaries are decided in this step. By process, floor plan in the physical design tool is the process of creating boxes which will house the submodules, memory macros, and standard cells on the silicon real estate. The floor plan is followed by the actual placement of the modules. Once, all the functional blocks/modules are placed, they are interconnected by a process called routing. Before this process, the power is distributed to all cells in the placed design and clock tree synthesis (CTS) is done. CTS ensures the clock is fed to all the timing elements in the design appropriately. Routing is a two-step process called global routing, and detailed routing. Global routing is the coarse routing where channels are created for routing which shows up the congestion if any, which is to be corrected by proper placement adjustments following which detailed routing is done. Every physical design flow is verified by extracting the netlist from the processed database and comparing it with the synthesized netlist which is the input to the physical design flow by a process called logical equivalence checks (LEC). Physical design is verified for signal integrity, [cross talk], antenna effects, and IR drop. Static timing analysis

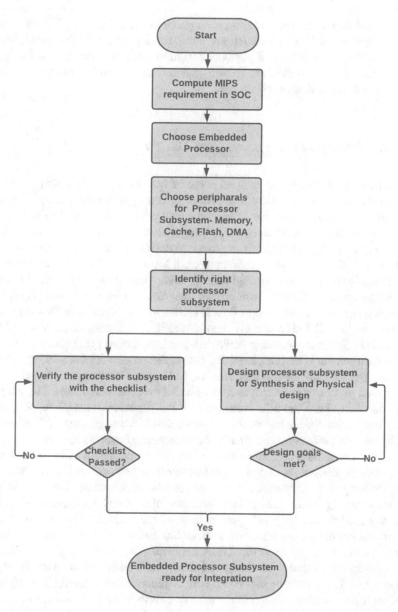

Fig. 2.7 Processor design flow

(STA) is done at every step of transformation of the design during physical design to ensure the timing goal is met. Once the physical design has passed all the verification goals, the file can be written out as a library file and the GDS II file formats. The library (lib) file of the design is written out if it must be integrated further with other design submodules for SoC design. As shown in Fig. 2.7, there is a parallel flow of activities during each phase of the design for different cores, like design verification by simulations, static timing analysis, DFT simulations, logic equivalence checks, and physical design verification, which must be completed satisfactorily before the design is taken up for further integration into the SoC at suitable design milestones.

2.26.3 Processor Subsystem Core Design

Embedded processors are an integral part of any system on chip design. In complex SoCs, there can be hundreds of processor cores performing general-purpose control functions and a number of special signal processing cores. Typically, processor subsystem cores are licenced or bought on royalty terms as soft or hard IP cores, unless the design center is in processor design. To ease the process of integration into different system architectures, a lot of flexibility on the processor cores is offered in terms of configurations. It is essential to arrive at the right set of configurations of the processor core to integrate into the SoC design. A typical processor subsystem core design flow is shown in Fig. 2.7. Processor subsystem core design in SoC design starts with assessing the processing power required for the system. This is expressed in MIPS (million instructions per second). Once the MIPS requirement is derived, available embedded processors from different vendors are assessed against this requirement, and options are compared to select the best suited processor and subsystem core based on other parameters. The parameters considered are the area consumed by the processor, customization needs while integrating, power consumed, software development platform, RTOS availability on the processor, and finally the commercial aspects like cost, loyalty terms, etc. Once the processor is chosen, supporting peripherals like Level 1 and Level 2 cache options, boot options, debug interface protocols, network interconnect supports, etc. are decided based on the SoC architecture. Selection of the processor configuration is based on modelling the typical application scenarios and, to an extent the designer's past integration experience. Major parameters in the processor configuration include address/data bus width, instruction/data cache sizes, peripheral subsystems like DMA controller, bus modules like AHB/APB bus master/slave, number of timers required, and number of interrupt lines, to name a few. The processor subsystem core is generated with the right configuration parameters and is verified in the standard verification environment provided by the vendors for confirming the claims on the performance and processing capabilities. Processor subsystem core can be soft core or hard core, which is interfaced with other blocks of SoC and the design process is continued. If the core is soft core, it is interfaced as a logic block, and if it is hard core, it is integrated during physical SoC design.

2.26.4   SoC Integrated Design Flow

The SoC design flow differs from the standard VLSI design flow only in integration flow. It can be considered a hybrid design flow where many core designs are in different stages of design abstraction get integrated. The design blocks/macros and IP cores to be integrated are made available in different types: soft core (RTL source code) or netlist, hard macro as liberty (LIB) file, or layout (GDS II) file. Analog/RF core design follows full-custom design flow, and processor subsystem core design is carried out using standard cell-based ASIC design flow to achieve high performance. These cores are integrated at different levels during a SoC design phase depending on the abstraction and the type of design. Figure 2.8 shows possible integration stages in SoC design.

At any design stage, an additional core gets integrated into SoC design database, and appropriate integrated verification has to be done to ensure that integrated design works as intended and design goals are met. SoC design continues after the integration of IP cores, with appropriate design constraint modifications and updated integrated verification on the revised design. The integrated design flow with the IP core integration is shown in Fig. 2.9.

2.27   EVM Design Development Flow

Simplest SoC validation platform is the circuit board with the SoC and all associated discrete components and modules which are used to validate the SoC for the specific features and the performance in the actual application scenario. EVM development flow begins as soon as the decision on the package is made, which typically is taken when the power-area number of IOs for the SoC is frozen. And in complex SoCs, which include multiple dies, the package design takes substantial efforts and time which need to be considered before the EVM development. The EVM design flow is shown in Fig. 2.10.

2.28   Software Development Flow

In earlier days, software development used to start after the hardware platform is available with the chip designed and fabricated. But with the availability of development boards with processor subsystems and high-density FPGAs, it is possible to develop the entire system on them and make them available for software teams to develop the SoC software much ahead of time during the SoC design cycle. Also,

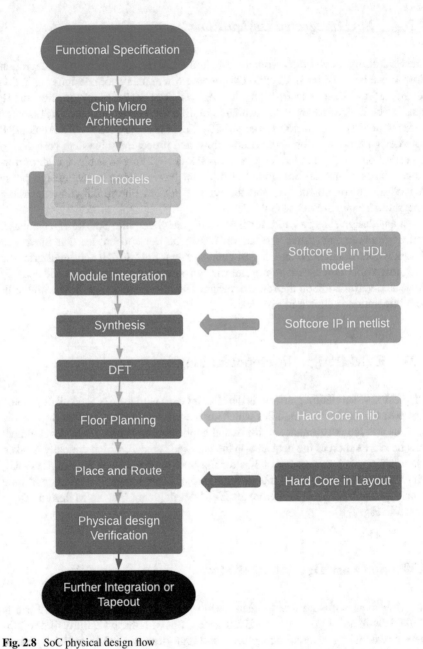

Fig. 2.8 SoC physical design flow

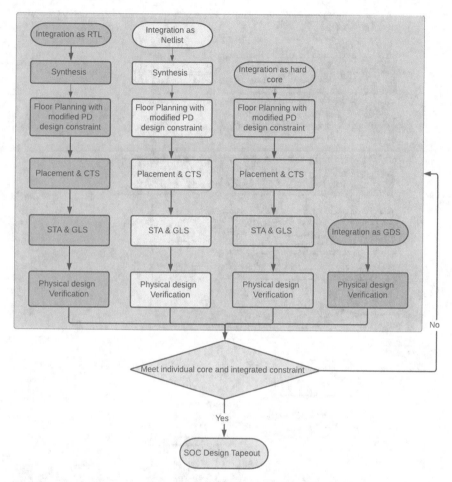

Fig. 2.9 SoC design flow with integration of cores at different levels of abstractions

the companies which provide processor cores provide development boards with processors and additional FPGAs for user logic design. They are used to validate the chip design by porting the design blocks into the FPGA. The development boards are used for software development in parallel to chip development. This helps to validate many of the design assumptions such as function latencies, interfaces, interrupt/DMA mechanism, etc. during SoC design. Intelligent algorithms are developed and validated in firmware to determine the configurations in real-time situations of the SoCs. Many times, selection of the right algorithm among many alternatives can prove to be the unique selling proposition of the SoC. The embedded firmware development flow is shown in Fig. 2.11.

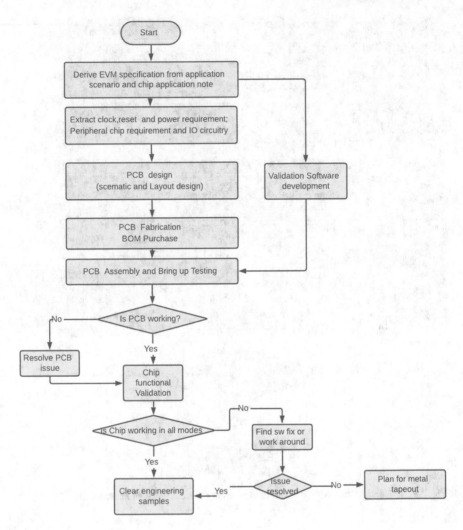

Fig. 2.10 EVM design flow

2.29   Product Integration Flow

Once the SoC design is validated on the EVM-based development platform, typically application notes are generated for SoC usage in various application scenarios for product design.

Fig. 2.11 Firmware
design flow

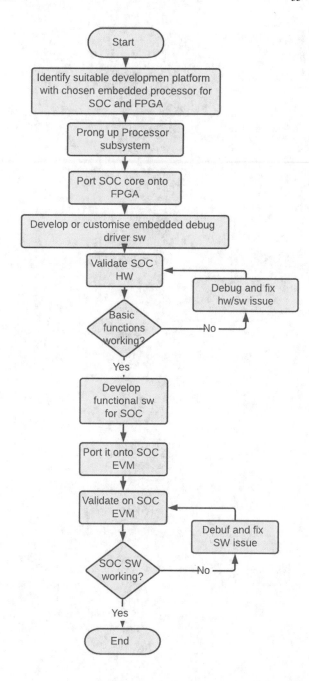

Chapter 3
SoC Constituents

3.1 SoC Constituents

A typical SoC consists of one or many processors and processor subsystems, on-chip memories, peripheral subsystems, standard communication cores, and peripheral device and memory controllers. An on-chip processor subsystem comprises single or multiple processor cores and standard peripheral bus bridges and interfaces. An on-chip memory consists of SRAMs or register arrays depending on the size. Major constituents of SoCs are application-specific functional blocks, protocol blocks, data processing blocks, and physical layer functions in communication processors or high-efficiency signal processing cores in multimedia SoCs; or a rule-based switching function in router SoCs. On-chip communication cores support communication with peer devices and make them interoperable. Some of the most commonly used communication cores are USB, UART, I2C, and SPI. Present-day SoCs also consist of high-performance mixed signal (analog and digital signal) processing blocks like ADCs/DACs, signal conditioning circuits, on-chip sensing functions for temperature and activity sensing, and functional blocks with radio frequency (RF) transceiver functions. Extra glue logic is added, which helps in housekeeping the data for processing or communication transfers, communication interfaces, and accelerators to support embedded firmware. Application-specific protocol functions and sensor/actuator interfaces with signal conditioning circuits and other control path modules like clock-reset circuitry, debug logic, DMAC, memory controllers, interrupt controllers, bus conversion modules, network interconnect modules, and DFT logic are typically found in SoCs.

© The Author(s), under exclusive license to Springer Nature Switzerland AG 2022   37
V. S. Chakravarthi, *A Practical Approach to VLSI System on Chip (SoC) Design*,
https://doi.org/10.1007/978-3-031-18363-8_3

3.1.1   Embedded Processor Subsystem for System on Chip

Typically, the main functional core in most of the SoCs, and embedded processors is the RISC processor which can be single or multiple instances depending on the processing power required for a particular application. As the technology allows integration of more and more cores on a chip, the SoC is being used for running hundreds of applications hence there are systems on chips as complex as embedding tens to hundreds of processors and peripherals on a chip. SoCs used for cloud server systems are examples of such complex systems. Embedded processors can be RISC processors or digital signal processors (DSP) or can be a combination of both in many numbers depending on the target applications. The ARM Cortex M4 embedded processor subsystem is one of the most popular processor subsystem cores in SoCs, are shown in Fig. 3.1a. As it can be seen, it consists of processor core, interrupt controllers, digital signal processing (DSP) core, floating point unit (FPU), memory protection unit, AMBA high-performance bus (AHB) lite interface, and a few of the debug interfaces like JTAG and serial wire.

Die photo of ARM 610 microcontroller SoC is shown in Fig. 3.1b. One can visualize the complexity and the density of a microcontroller SoC.

Choice of Embedded Processors for SoC

Selection of the on-chip processor and its subsystem is purely based on the processing needs of the system. Processing power is determined based on the application. The functionality is defined and classified based on the criticality of the system.

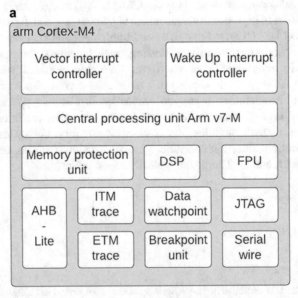

Fig. 3.1 (a) ARM Cortex M4 block diagram. (b) Die shot of ARM610 microprocessor. (Source: ARM 610 microprocessor; Courtesy: GEC Plessey Semiconductors)

b

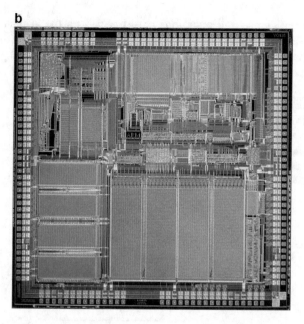

Fig. 3.1 (continued)

With the hardware-software partitioning of the functions, processing requirement on the chip are derived. Though there is no formal process of deriving the processing requirements, the typical activities followed to arrive at the requirements are the following:

- List the functions to be executed in the software after hardware-software partitioning for SoC.
- Classify them as functions which can be executed by general-purpose instructions and signal processing instructions (meaning the functions requiring math operations such as multiplication, division, filtering, etc.). General-purpose functions are mapped to RISC processors and signal processing functions to digital signal processors.

Embedded General-Purpose RISC Processors

- Classify the functions into real-time and multicycle operations.
- List all the processes in the functions in the multicycled operations.
- Map the processes of the functions to load, operate, and store instructions of the general-purpose RISC processors.
- Add all the instructions required to execute all the operations and multiply them with the average number of cycles per instructions, from which derive the number of instruction per second. This is the processing required for the operation. Many times, it will not be straightforward as it is said here. In such cases, such

functions, programs, and algorithms are modeled on the available processor development platforms to assess the number of read/write instructions and arithmetic/logic instructions required for the operations.
- Map the requirement to available processor datasheet parameters and compare them against each other.
- Choose the best suited processor core.
  The selection process is shown in Fig. 3.2.

Fig. 3.2 Selection process of embedded processor for the SoC

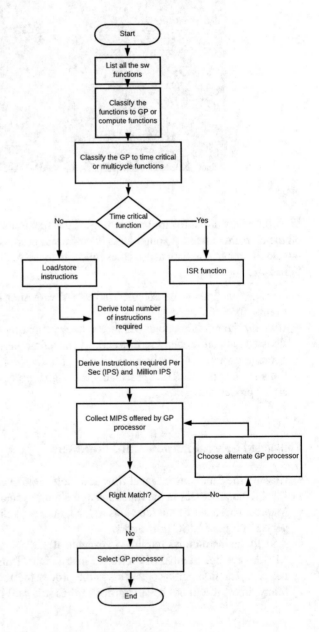

Case Study 3.1 To arrive at the MIPS requirement for packet processing in Ethernet packet of size 256 bytes

Structure of an Ethernet frame is shown in Fig. 3.3.

As shown in Fig. 3.3, a typical Ethernet frame contain a preamble, start frame delimiter (SFD), MAC header with destination and source addresses, Ethernet frame type, and the user data followed by frame check sequence (FCS). The two Ethernet frames are separated by interframe gap (IFG) which is the known idle pattern. To find the MIPS of the processor which has to process such frames, it is essential to know the frame structure. Please note that the frame size can be of any size between 64 bytes and 1864 bytes. Ethernet also supports jumbo frames of larger than size of 1864 bytes. For all the size of the frames, it is essential to derive the data throughput with technology overhead. All assumptions are considered while arriving at MIPS requirement of the processor is in Table 3.1.

The user data throughput is defined as how much of user data (payload) can be transmitted excluding technology overheads like preamble, header, and FCS per second.

Number of devices the system supports: 128.

Part of frame to be read to process it: 40 bytes (header part of the packet only).

Number of reads/writes required for processing 40 bytes: 10 (depends on processor data bus width).

Number of reads to be done on configuration and device detection: 128.

Number of compare operation to be done to detect the device: 128.

Number of writes: 5.

Total processing per frame: $10 + 128 + 128 + 5 = 271$ operations.

Number of frames per second: transmit/receive rate/(frame size in bytes*8).
$= 700,000,000/(128*8) = 683593.75$.

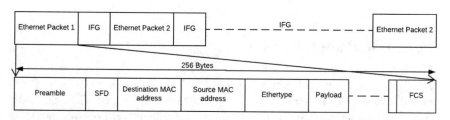

Fig. 3.3 Ethernet frame format

Table 3.1 Assumptions regarding Ethernet frame transmission

Frame part	Value	Unit	Remarks
Preamble	2	uS	Time to transmit
Physical layer header	582	nSec	
Guard interval	36.4	nSec	
Transmit/receive rate	700,000,000	Bits per sec	Rate of transmission
MAC header	40	Bytes	Field size

Number of operations needed to process frames of size 128 bytes per second: 185253906.

Number of millions of operations (MIPS) needed per second: 185253906/1000000 = 185.26MIPS rounded to 186.

Some amount of MIPS required to manage the connected devices and link management which can be assumed as 15% which will be 0.15 *frame processing MIPS.

Total MIPS required: 193 + 0.15*186 = 213.9 MIPS rounded to 214.

But note that fixed size frame is considered for computation and in practice, the Ethernet frame can be of any size between 64 to 1836 bytes, and it is customary to assume 40% more MIPS to accommodate the random frame sizes and other overheads.

MIPS required for this SoC = 214 + 0.4*214 = 299.6 rounded to 300. Any embedded processor with more than 300 MIPS will be good enough to process single port Ethernet frame processing SoC. However, if the SoC has to process multiple ports, then the MIPS required has to be multiplied by the number of ports.

The intention of this case study is to give the rationale behind choosing a processor based on MIPS and not the accurate one. It is to be noted that the processor selection as shown in Fig. 3.2 is based on the technical feasibility of SoC. Many times, the choice of processor depends on other factors like the customization required to integrate, power consumption, area of the core, etc. Other factors affecting the processor decision are development support for embedded software and the support software such as compilers, and operating systems (RTOS). Sometimes, it can become a business decision. The processor IPs are sourced with licence and call for royalties when SoCs are manufactured in large quantities. Processor cores are also available as hard micros for SoC integration which reduces design time drastically. Processor cores such as RISC V are available free of cost for SoC designs, which promote the democratisation of SoC development.

3.1.2   DSP Processors

Use of SoCs in Internet of things (IOT) and multimedia applications necessitates the integration of many real-time signal processing functions on VLSI chips. An example is a multisensor SoC where the signal conditioning feature samples a large number of real-time physical parameters periodically, averages them over time, and digitally filters the noise to derive meaningful data. Most of the communication protocols SoC features demand digital signal processing for baseband level data processing. Also, there are exclusive digital signal processor-based SoCs which are optimised in terms of area and power to be able to integrate on the chip for these applications. In most applications, the SoCs use DSP processors for external interface to samples and signal/data front-end processing to derive meaningful information, which are further processed by RISC processors on chip for data interpretation and analysis. This can be seen in all multimedia and communication SoCs.

3.2 Issues of Hw-Sw Co-Design

The complexity of SoC design increases considerably with the requirement of hw-sw co-design to get time to market advantage. The in-system programming (ISP) adopted for software development during SoC design poses a lot of challenges for achieving high performance. The ISP is primarily used for data computing and control systems. The need for application-specific, retargetable compilers and embedded assembly level programming makes it more complex. It is necessary to decide on the need for software accelerators and hardware accelerators/coprocessors at the early stages so that targeted high system performance can be met. Most of the time, the software development time exceeds the hardware integration time for embedded processors in SoC development. Sometimes, in safety-critical applications, it is necessary to plug in safety functions in a chip which are of more importance than performance. In such cases, computer-aided compilers are generated and used. Decisions like this, require system level knowledge, experience to foresee the issues and add work around plug-ins in the SoC.

3.2.1 Processor Subsystems

The processing engine is the heart of the system but for it to run an application, just the processor alone is not enough; it needs many peripherals. Some examples of peripherals are on-chip flash modules for booting; internal SRAM memories, for storing the data and program codes; cache, or scratchpads; UART/JTAG for debugging program execution; DMACs; interrupt controllers; etc. Many SoC applications need large expandable memories which require robust high-density memory controllers. All these togather constitute the processor subsystems. A typical processor subsystem used in IOT applications is shown in Fig. 3.4. The processor subsystem in a SoC includes a processor core interfaced with peripherals which are either proprietary or generic. The processor core is interfaced to other intellectual property (IP) cores and standard functional blocks such as DMA, memory controller, and radio controller blocks on a standard high-performance buses such as the AHB expansion port, and peripheral subsystems such as the ADC, DAC, and I2C interfaced through low-performance peripheral bus such as APB interface, on-chip and off-chip memories interfaced to a memory controller, embedded flash connected to flash controller, and flash cache controller.

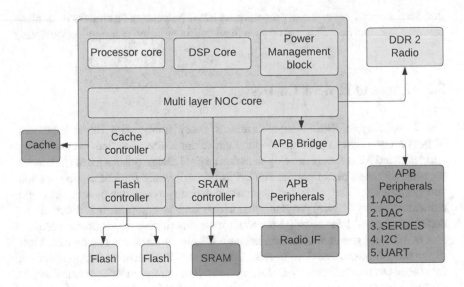

Fig. 3.4 Processor subsystem

3.2.2 Processor Configuration Tools

Considering the complexity of SoC designs, the processor vendors offer processor system configurable tools to decide on the suitable configurations of processor cores. The processor configuration tool is used to generate various software development platforms and custom tool kit for the selected processor subsystem. This helps to model on-chip processors of desired configurations and automatically generate the corresponding toolkits for the processor hardware/software co-design. This also generates the test setup for subsystem verification and co-verification environments, including the instruction set simulator. Embedding a processor subsystem requires designers to work in two fields: hardware development of the processor architecture and software toolchain development for the compiler, assembler, linker, simulator, and debugger. Both use the software simulator to profile data to identify hotspots and bottlenecks in the instruction set, analyze the performance of an algorithm, and determine the required size of memory and registers. In addition to architecture exploration, the tools provide ways to generate hardware design files, including RTL files, and other physical design files and system level descriptions using modelling languages like MATLAB or System C. The flow in Fig. 3.5 shows the choice of parameters in a typical processor subsystem configuration tool.

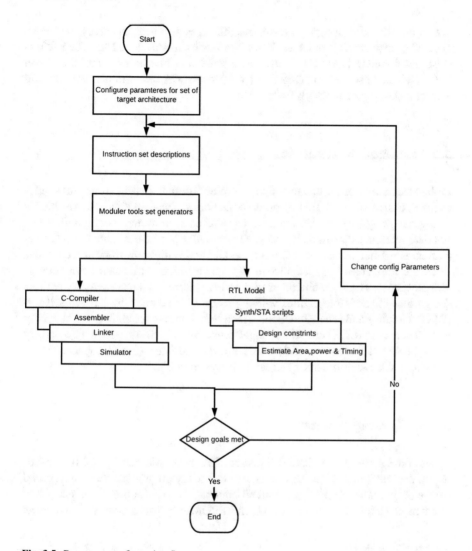

Fig. 3.5 Processor configuration flow

3.2.3 Processor Development Boards

To reduce the risk involved in fabricating the complex SoCs, the designs are first validated on the development platforms that have hard processor chips equivalent to the processor core used in them. The development board has a high-density field-programmable gate array (FPGA) to which the SoC design is ported. This helps in validating the SoC modules and serves as a development platform for system soft-ware design. Processor core developers also provide development boards based on processor chips or as hard macros in FPGA. The development board is used for evaluating the SoC design performance, like speed, power, accuracy, and cost. Typically, software drivers for SoC interface blocks are developed and validated on these development boards. It is used as a validation platform for custom IPs on FPGA, which must work in tandem with the processor. A few examples are Juno and Neoverse development boards from ARM.

3.3 Embedded Memories

Embedded memories are an inevitable part of SoC design. In fact, around 40 to 60% of the SoC area is constituted by on-chip memories. They are either SRAM blocks or register arrays. Memories are used to store data, or system configurations, or standard reference data in the systems. On-chip SRAMs are available for SoC integration as memory arrays of configurable sizes with different rows and columns. There are companies that specialize in high-quality, high-performance memories of different types. These are silicon proven and are offered as a macro library with all design files for SoC integration. Memory macros also come with built-in self-test (BIST) circuitry and repair functions which help improve testability and high chip yield. There are SRAM cells with a single transistors, which are used in high-density SoCs. SRAM cell which is widely used has a six-transistors (Six T). Typical SRAM cell with a six transistor (Six T) structure is shown in Fig. 3.6.

3.3.1 Types of Memories

Types of memories used in SoC designs are SRAMs, ROMs, and EPROMs, depending on the requirement. The EPROMs are electrically programmable with a special device programmer. Typically, the small boot vector code for processor subsystem or the reset vector can be loaded into such EPROMs as a part of power-on sequence.

Fig. 3.6 6 T SRAM cell structure

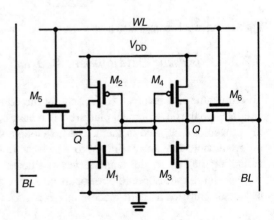

ROM has to be loaded with the initialization data from the fabrication facility itself during automatic test screening. So, when the SoC design contains ROM, the vector file must be submitted to the fabrication/test house. Memory vendors offer memories as macros of different types that are highly optimized for size, power, and access times that are silicon proven. Memory macros are generated as register files of register arrays, single-port SRAMs (SPSRAM), dual-port SRAMs (DPSRAM), and SRAMs/DPRAMs. There are special memory macros with redundant storage that are called repairable memories. The redundant part of memory macros is generally not used unless some of the memory cells have resulted in faulty ones. Special memory macros also come with error checking code (ECC) logic which is used as repairable memories. ECC RAM monitors data as it is processed by the system, using a method known as parity checking, and if there are errors, they identify and correct them.

3.3.2 Choice of Memories

On-chip RAMs have different types of faults than those in logic functions. To avoid them, special care has to be taken during chip design at the cost of increased silicon area. Some of the most commonly used special design techniques are built-in self-test (BIST) logic, repairable logic, etc. The choice of the type of on-chip memory is based on the criticality of the memory content, access timing requirements, and overheads affordable on silicon real estate.

3.3.3 Memory Compiler and Compiled Memories

As mentioned earlier, the on-chip memories for SoCs, are available as macros of different types. Memory macros are proven cores in a process technology. Memory macros are optimized for PPA to suit SoC applications. They are size and structurally organised in terms of number of rows and columns (R x W) and placement orientation on the layout configurable. They are configured by a tool called memory compiler. These compilers generate design files for a chosen macro configuration. Based on user preference, the memory compiler can write out different design files required for SoC integration and design. The macro design files include front-end (HDL) models, test benches, test scenarios, and physical design files. Memory compilers also generate special memory macros such as repairable memories and those with advanced power management modes, such as light sleep, deep sleep, and shut down. They can generate macros with high-speed sense amplifiers, fast clocking, and fast bit line recovery to achieve the high speed required by today's high-performance applications. In summary, the memory compiler:

1. Generates memory instances that include all the necessary logic to facilitate at-speed built-in self-test (BIST), ECC, and redundancy for repair for user configuration.
2. Generates memory models with different aspect ratios, test benches, liberty files, GDS II, and LEF plus many other views in one concise database.
3. Generates high-performance memories in terms of access times or high yield factors by the selection of process-sigma characterization and read-write margin settings.
4. Automates the process of generating macro design files for SoC design integration using standard EDA tools.
5. Has user-friendly graphical user interface (GUI) to generate large number of memory macros in batch mode with fast run time.
6. Generates the fully encrypted and protected physical design files as these are characterized for set performance.
7. Generates PDF datasheets corresponding to the macros generated.
8. Operates independently from EDA tools.
9. Generates user design guides with training and tutorials for SoC integration.
10. Generates real-time instance-based characterization.

Typical memory compiler architecture is shown in Fig. 3.7.
Intel's 22 nm technology SRAM memory die is shown in Fig. 3.8.

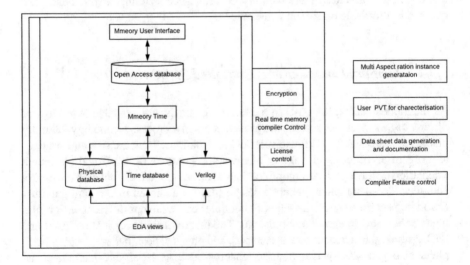

Fig. 3.7 Memory compiler architecture

Fig. 3.8 Intel's 22 nm SRAM memory wafer. (Source: SemiconDr blog; Photo courtesy: Intel)

3.4   Protocol Blocks

System-specific functions are designed by a single or set of functional blocks that execute tasks in proper coordination in a well-coordinated manner. It can be a subsystem in the SoC that executes a protocol function. A protocol is a series of processes, involving two or more functional units in a system, designed to accomplish a function. The typical characteristics of the protocol are the following:

1. All blocks are part of the protocol and execute their identified predetermined function.
2. All blocks coordinate and perform in tandem to execute a function.
3. Protocol function must be unambiguous.
4. It must be complete with all conditions clearly defined.

Protocol can be technology defined, or application dependent, or process dependent. Examples of technology-dependent protocols are the Bluetooth protocol, WLAN protocol, and Ethernet protocol as defined by the respective technology standards. These standards are defined by professional bodies like ITU-T, IEEE, etc., and are accepted widely by the developer communities and help in interoperability. An example of process-dependent protocol is the cryptographic protocol. This protocol is used to avoid hacking or data misuse. Some of the protocol examples are shown in Fig. 3.9. For easy understanding and compliance, protocols are represented by a state diagram 3.9a message sequence diagram 3.9b, and a dataflow diagram 3.9c.

The protocol block is designed to be intelligent enough to know the configurations and respond to the contexts defined in standard protocols. Figure 3.10 shows an example of the IEEE 802.3 standard-based 10/100Mbps media-independent

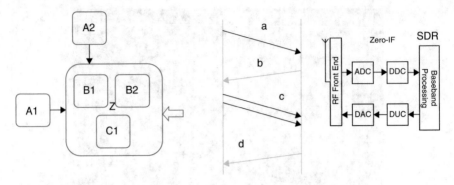

Fig. 3.9 (**a–c**) Protocol examples

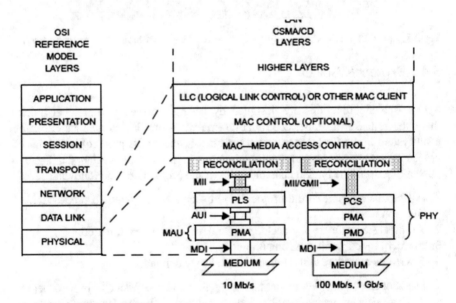

AUI = ATTACHMENT UNIT INTERFACE	PCS = PHYSICAL CODING SUBLAYER
GMII = GIGABIT MEDIA INDEPENDENT INTERFACE	PHY = PHYSICAL LAYER DEVICE
MAU = MEDIUM ATTACHMENT UNIT	PLS = PHYSICAL LAYER SIGNALING
MDI = MEDIUM DEPENDENT INTERFACE	PMA = PHYSICAL MEDIUM ATTACHMENT
MII = MEDIA INDEPENDENT INTERFACE	PMD = PHYSICAL MEDIUM DEPENDENT

Fig. 3.10 IEEE802.3-based 10/100Mbps MII protocol. (Courtesy: IEEE)

interface (MII) protocol as applied in the OSI reference model (detailed in the latter part of the chapter).

In Fig. 3.10, the protocol block includes a physical layer design function. This includes managing transmission media such as air in wireless and wires in wired technologies. The protocol functional block takes care of signal level and strength, noise filtering, configuring the air/connector interface, and necessary signal processing. The physical medium interface supports signal processing with physical layer coding sublayer (PCS) functions, such as encoding/decoding, scrambler/descrambler, and 3B/4B code converter. The physical layer block is interfaced to the media access controller (MAC), which is the data link controller block and logic link control functional block in the system. The details of the functionality are out of the scope of the book.

3.5   Mixed Signal Blocks

Designers can now integrate mixed mode signal processing blocks in SoC, thanks to advancement in design automation tools. The SoC design methodology permits interfacing analog and digital signal processing blocks, thereby reducing the bill of materials (BOM) of the product. Examples of the mixed mode blocks are data converters, transceivers, etc. There are two types of data converters: analog-to-digital converters (ADCs) and digital-to-analog converters (DACs). These enable you to connect the SoC to sense, process, and monitor physical parameters using sensors and transducers, like such as microphones, speakers, cameras, and accelerometers. The mixed signal mode blocks are interfaced as per the technology standards like Wi-Fi, Bluetooth, MoCA, PLL, or proprietary interfaces using some of the transducers: temperature, accelerometers, and pressure and sound sensors as required by the applications. An example of data converter is shown in Fig. 3.11. There are design companies that exclusively design these mixed signal IP cores and offer them for SoC integration. Analog and mixed signal design methodologies are more complex and involve more manual processes compared to digital design methodologies.

3.6   Radio Frequency (RF) Control Blocks

Advancement in process technologies has enabled the realization of intermediate and radio frequency signal processing on chip. Signal processing functions like the modulation and demodulation, filtering at intermediate frequencies, and realization

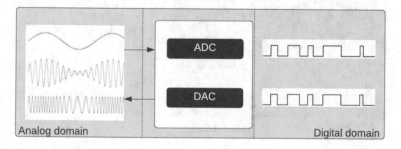

Fig. 3.11 Data converter for SoC integration

of antenna structures on chip are possible to realize in a CMOS-compatible fabrication process called the RF-CMOS process. Modern communication technology operates with high data rates of the order of gigabits per second. These adopt complex signal modulation schemes applied to data transmitted on high bandwidth of the order of 80 MHz, communication channels. This has resulted from aggregation of channels, complex multi-antenna array architectures, and interchannel noise cancellation techniques. From the baseband perspective, the multi-antenna results in multiple data streams processed, requiring multi-analog interface modules. A typical WLAN 802.11 ac SoC implementation uses more than two data stream transmissions with antenna array configurations. Hence, in most high-performance communication processors, TV processing SoCs, IF, and RF transceivers are inevitable.

3.7   Analog Blocks

Typically, signal conditioning is an analog function. It is carried out in analog blocks which are integrated into the SoC as macro. They are developed or bought as third-party intellectual property macros for integration into SoC during the physical design stage. Analog blocks are designed using custom layouts which are hand-crafted and validated mostly by test chips. One such example is a phase-locked loop (PLL) block, which is used to generate fixed and variable frequency clocks on a chip. An example of PLL is shown in Fig. 3.12.

In the full-custom design flow, design is done by drawing the schematic circuit using circuit elements like transistors, capacitors, resistors, and inductors, which are interconnected using the schematic editor tool. Most EDA tools come with schematic editors for design input. Circuit simulation for analog blocks is done at the transistor level using circuit simulator tool such as Simulation Program with Integrated circuit Emphasis (SPICE). The standard cell library cells for automatic cell-based design flow are designed using a custom design flow. This design methodology is most commonly used for digital designs.

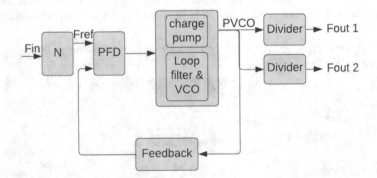

Fig. 3.12 PLL block diagram

3.8 Third-Party IP Cores

It is quite common that apart from specialized SoC constituents explained, it is necessary that it contains standard interface IC cores like UART, USB and SPI to expand and interface with external ICs to enhance the capability. These interface cores are called intellectual property (IP) cores and are bought from third-party vendors, on licence and royalty terms. SoC IP cores are pre-verified and pre-validated functional blocks ready to be integrated into SoC. The IP cores are purchased as soft cores or hard cores depending on the target technology and customization required for integration. Soft IP cores come with design files, test benches and synthesis setups with design constraints with which they have to be synthesized. When IPs are bought as hard macros, no customization is possible.

3.9 System Software

System software is the integral part of a system on a chip in today's world. The software can be classified in many ways.

3.9.1 OSI System Model

The communication system layers are classified depending on the functions they perform and how closely they interact with either the hardware or the application that interacts with the user. Figure 3.13 shows the most common OSI model of the system layers for network systems as defined by the International Organization for Standardization (ISO). The same model can be used to explain other systems on the chip by collapsing some of the layers. System on chip designs typically identify all time-critical functions of mostly layers one, two, and three, collapsing them for implementation on chip in total or as an accelerator engine for firmware implementation. Fourth and above layers are implemented on general-purpose processor or computational systems which interact with the SoC hardware.

Brief introduction of OSI model is given in this section.

Physical Layer (Layer 1)

The physical layer constitutes the physical layer signal processing functions along with physical link control functions like signal boosting, modulation and demodulation, received signal detection, carrier detection, link establishment and maintaining functions, encoding and decoding, clock recovery functions, and detecting valid physical layer packets and passing them onto the data link layer.

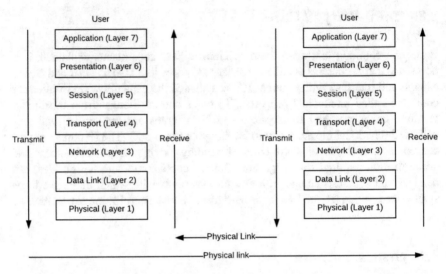

Fig. 3.13 OSI model of system layers and their interactions

Data Link Layer (Layer 2)

Data link layer functions include peer-to-peer data transfer functions and error detection and correction functions. It includes two sublayers: the data link layer and logic link layer sub-functions.

Network Layer (Layer 3)

This layer includes functions of networking and routing to different nodes and interfaces by detecting the source and destinations and applying certain accepted rules. Also, this layer manages the packet routing functions to different nodes and even routers.

Transport Layer (Layer 4)

This layer is responsible for coordinating data transfer from host to system, deciding the data rate, bandwidth and throughput.

Session Layer (Layer 5)

When peer-to-peer link is set up, the session has to be set up for data transfer between the two devices. Session layer sets up the session for data transfers and terminates it after completion.

Presentation Layer (Layer 6)

The presentation layer represents the preparation or translation of data from application format to network format, or from network formatting to application format data. In other words, the layer "presents" data for the application or the network. A good example of this is the encryption and decryption of data for secure transmission—this happens at the presentation layer.

Application Layer (Layer 7)

Application layer is a user interface. It accepts data from the user for transmission or further processing and communication. This layer corresponds to users.

3.10 GAMP Classification of Software

System layer classification is also done according to the definition of good automated manufacturing practise (GAMP), a technical subcommittee of International Society for Pharmaceutical Engineering (ISPE). According to this, the hardware, firmware, device driver, middleware software and newly added cloud are all system layers. The software which interacts with the user is also termed "human ware". Figure 3.14 shows the system layers and their interactions. GAMP classification which include risk assessment and traceability as best practise guidelines was defined for pharmaceutical systems but is now widely used in all other domains. A brief description of the classification layers follows.

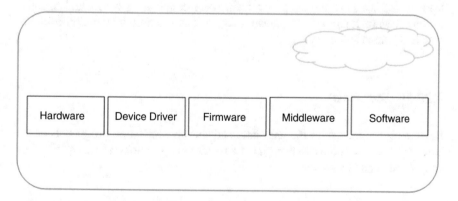

Fig. 3.14 System layers and their interactions

3.10.1   Hardware

Hardware includes SoC and supporting peripherals, which are the main parts of the system or the solution.

3.10.2   Device Driver

The device driver is closely related to the hardware and is used to control the hardware functions. Examples of device drivers are display controllers, keypad controllers, interface controllers like I2C master/slave drivers, Bluetooth module drivers, etc. It can reside in flash memory.

3.10.3   Firmware

When system is partitioned into hardware-software (hw-sw), where the software part of the program complements and completes the function in association with the hardware. It includes algorithms, protocol interpretations, decisions-making based on the various events, and the state of hardware. It typically resides in ROM, EPROM, or flash memory. Bare metal (which directly works with hardware without an operating system) or based on real-time operating system.

3.10.4   Middleware

Middleware is a software that connects firmware or operating system and an application software. It particularly manages complex transactions with multiple distributed application software.

3.10.5   Software

Remaining program with the user interface and an application program is called software. It converts the messages and transactions and deciphers in a way that they can be consumed by the user.

3.10.6   *Cloud*

Cloud part of the system structures the large data generated by the system, stores, processes, and analyzes reliably and securely for the user/users who are entitled to access them. It is the shared resource where the user can selectively access his portion of the data for consumption.

As above classifications enable the correct development of a complex system, with advancement in chip technology, most of the system gets implemented in chip or memory chip or processor chip or server/storage system on a chip and packaged as a solution.

3.11   Design-Specific Blocks

Apart from the system functional blocks, design methodology and technology-specific blocks for achieving the yield and reliability of devices are required for the SoC. They are on-chip clock (OCC) generator block; power management block; on-chip process, voltage, and temperature (PVT) sensors (for chip health monitoring); and design for testability (DFT) logic which guarantee the reliability, safety, and testability of the SoC.

Chapter 4
VLSI Logic Design and HDL

4.1 SoC Design Concepts

Basic logic functions are combinational and sequential circuits, but the majority of the SoC designs are sequential designs as it is easy to represent system functionality by data/control path architectures. This is easily done by defining their different states spread across time. Data path architectures of system definition can be extended to most of the subsystems if their functionality can be classified as a finite number of states. This requires identifying the system functionality as small logic partitions and realizing them as fundamental logic circuits. Some of the SoC design concepts essential for defining the functional architecture of a system are as follows:

- Logic design fundamentals.
- Synchronous sequential functional blocks.
- Speed matching.
- System state machines.
- NOC architecture.
- System modes.
- Hardware accelerators.

4.1.1 Logic Design Fundamentals

System design concepts are based essentially on logic design fundamentals. In this section, we review the design concepts of logic design. Logic circuit functions are classified as sequential and combinational logic. Circuits which require a clock for its operation are called sequential logic circuits. Typically, they either store the data for processing or involve memory. Examples of sequential circuits are timers, counters, multipliers' register arrays, etc. Subblocks are designed using combinational

© The Author(s), under exclusive license to Springer Nature Switzerland AG 2022
V. S. Chakravarthi, *A Practical Approach to VLSI System on Chip (SoC) Design*,
https://doi.org/10.1007/978-3-031-18363-8_4

and sequential logic using a clock as a main block control signal. Such blocks are synchronous logic blocks. A clock signal helps to time every event in the dataflow architecture and control the data movement during data processing. Clock is fed to all sequential timing elements such as flip-flops and latches. These are design elements used as storage elements in the data path. Logic functions are realized using combinational logic gates and sequential cells called standard cells. NAND, NOR, XOR, INV, and buffer gates are combinational cells and D flip-flops and latches are sequential cells called "standard cells". Some of the complex cells, such as adder, subtractor, encoder, decoder, and multiplexer, are designed using primitive gates to form complex design elements. The design method which is used to model the design with a dataflow architecture is mainly register transfer level (RTL) design. Any logic design can be represented in RTL design by representing the behaviour and structure or in terms of algorithms, which are essential components of subsystem design. All designs that use the clock as the main control signal are synchronous designs. All other control signals in such designs are synchronized to the main system clock.

As previously stated, any logic function is represented as RTL design using hardware description language (HDL). RTL designs are converted to design netlist by the process called synthesis using the EDA synthesis tool. The synthesizer tool optimizes the logic functions, based on the cost function which could be area or time as desired by the designer. Logic optimization techniques involve two-level logic optimization, signal reordering, logic sharing, etc.

4.1.2   System Clock and Clock Domains

A set of logic design blocks operating on a single clock is called clock domain. Complex SoCs will have hundreds of clock inputs driving different parts of a logic circuit and, accordingly, several clock domains. A clock is called the primary clock if it is the output of the clock generating circuit called the clock source. A clock source for a SoC will typically be a phase-locked loop (PLL) circuit. Clock is a derived clock if it is generated from the primary clock by dividing or multiplying it internally by a constant or by introducing phase delays. As there are several clock domains in the SoC, the data signals traverse across subblocks crossing different clock domains. The clocks in different domains can be of the same frequency and different phases or different frequency and phases. Because most logic designs use a clock to latch data, it is critical to take special care to generate data with generated clock so that they are correctly latched. The generating clock is also called the launch clock. When asynchronous signals cross clock domains, it is necessary to identify the critical data and control signals and synchronize them to the receiving clock to ensure that they are stable for at least one clock cycle of the receiving clock. It is better to keep the data signal stable for multiple clock cycles of the receiving domain. An example is illustrated in Fig. 4.1.

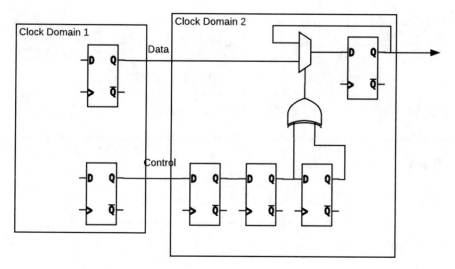

Fig. 4.1 Clock domain crossover

4.1.3 Asynchronous and Synchronous Resets

System must be deterministic in terms of logic states for stability and reliable operations. To make the system deterministic, it is necessary to initialize the circuits to a
known state. This is done by a reset signal. This signal can be of a pulse, or level,
signal. A reset is an external signal, that resets all the logic states to a known default
state or pre-defined state. This signal can be asynchronous from an external switch
or synchronized to the system clock to make it a synchronous reset event.

Metastability

Badly designed circuit can get into a condition where the signals can settle to an
intermediate value between logic 0 and logic 1. These signal states are nondetermin-
istic in terms of logic 1 or 0. This is called metastable state. When this happens, the
logic circuit in the system may not return to a stable state and can get stuck in
a metastable state, leading to fatal system errors. This will happen when the proper
timing requirements of the sequential design elements are not met as required. The
timing elements such as flip-flops and latches are characterized by setup and hold
time requirements for correct operation. Two main timing requirements are setup
time and hold time. The setup time of a flip-flop is the time duration for which the
data should attain a stable logic value (1 or 0) before the active clock edge. The hold
time is the time for which data should remain stable after the active clock edge. The
data arrival time at the inputs of a flip-flop or latch must satisfy setup and hold times
for proper operation. If this is not met, the circuit can enter a metastable state and,
most of the times will not return to stable state. This is avoided by meeting the proper

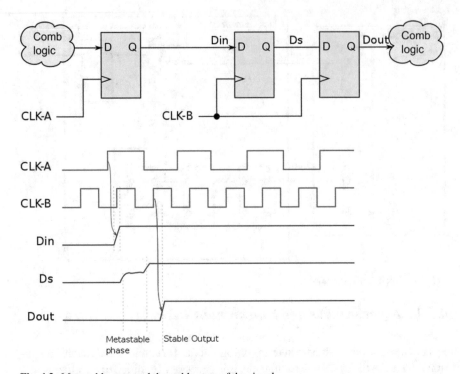

Fig. 4.2 Metastable state and the stable state of the signal

timing specifications of design elements. If the signals are asynchronous to the capturing clock, a technique called double synchronization is adopted. Here, the data inputs are synchronized with the clock by passing them through two or more flip-flops, thus giving enough time for them to settle down to a stable state before they are used for processing. Figure 4.2 shows the logic path in metastable and stable states.

Standard Cells and Compiled Logic Blocks

Vendors provide commonly used circuit blocks as a standard cell library. Standard cells are predesigned, pre-validated for functionality, and pre-characterized. A typical cell library contains all cells corresponding to generic standard cells of both straight and inverse functions such as NAND, AND, NOR, OR, XOR, XNOR, INV, and buffers that are used in the process of synthesis as a minimum set of cells. It also contains mega cells which surpass the complexity of standard gates like AND-OR-INVERT (AOI), clock buffer (two cascaded inverters), INVERT-OR-AND (IOA), and some complex functions such as adders, multipliers, multiplexers, encoders, and decoders, which are most commonly used in SoC designs. The library of pads contains different types of PAD cells such as input pads, output pads, and bidirectional pad cells with drivers of different drive strengths. The pad cells are targeted at CMOS-compatible technologies or mix technologies like Bi-CMOS processes. These are optimized for power, area, and timing and are used in automatic cell-based SoC designs. Similarly, memory macros of various types. SPRAMs of single port static RAM and DPRAMs of dual port RAMs, SPRF of single port register files (SPRF), and DP register files (DPRF) are examples. Different configurations of the memories of different sizes can be generated using memory compilers for SoC designs. Most of the semiconductor companies like Intel, Texas Instruments, and IBM own their own fabrication units where they fabricate the SoCs designed by them. Apart from this, there are also other contract fabrication companies like TSMC, GlobalFoundries, etc. that accept SoC designs from fabless design houses. This enables the fabless design centers to provide design as a service and realize various system on chip (SoC) designs without the need for owning fabrication facility.

Hard and Soft Macros

Macros are VLSI designs that are ready to be used in SoC designs. Macros are available as soft cores or hard cores. They are available on licence or royalty terms for reuse in SoC designs. Soft macro is a core with source code in HDL behavioural module to be integrated at the front-end or logic design stage before synthesis. This allows the designer to customize the core to make it suitable for SoC integration. Synthesis is carried out after integration. Hard macro is a core which is integrated at the physical design stage. The hard macros cannot be edited or upgraded. Processor and subsystem cores and interface cores are available as hard and soft macros for SoC designs. Some of the examples of processor macros are Cortex M3/M4 and advanced cores from ARM, ARC core from Synopsys, MIPS core, standard interface cores such as PCI express, USB, UART cores, high-performance interconnect/ interface blocks like AHB master-slave cores, AHB-APB bridge, and AXI interconnect cores from ARM.

Data Buffers and Buffer Managers

In systems, data is always stored for processing or forwarding in the on-chip memory or external memory interfaces. On-chip memories are arrays of registers such as SRAMs, and external memories are SDRAM, DDR, etc. These memories are managed for efficient storage and retrieval by efficient access when required in the systems. This is done in memory or buffer managers, or controllers. The memory managers or controllers adopt techniques like linked lists and queuing the data for efficiency in data access. Different types of memory controllers are available as IP cores for SoC integration. They boost overall system performance in SoC.

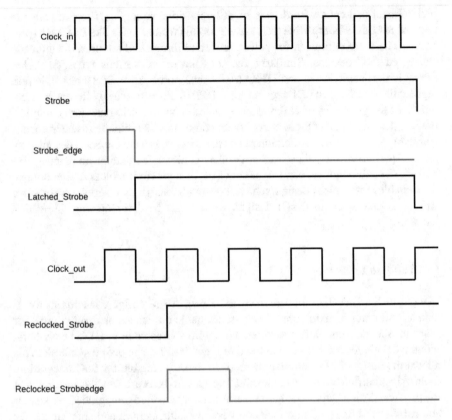

Fig. 4.3 Design assertion example

Design Assertions

Design assertions are the design events generated in SoC designs to check the temporal relationship of synchronous signals for the correct functioning of the module. The assertions in designs are tracked by the test bench checker module. The events which are sure to happen for the correct functionality are monitored constantly during design verification. A common example is when the signals from faster clock domains cross to slower clock domains. Figure 4.3 shows a timing example. In this example, by design, the signal *Reclocked_Strobeedge* must be set if the *Strobe-edge* signal is set. The design assertion is added in the SoC design for monitoring, but not the setting of *Reclocked_Strobeedge* when *Strobe_edge* is set, which indicates the design issue, which can be debugged during verification. This is an important technique to make SoC design verifiable.

4.1.4 Synchronous Sequential Functional Blocks

Control path in SoC designs are implemented using finite-state machines (FSMs). Hence, FSMs are unavoidable in digital system designs. FSM-based subsystem functions occur in a particular sequence and are repeatable if subjected to the same set of input conditions. The system states are distinct and are stored in on-chip memory. FSMs require the system states to be encoded and stored. There are two types of FSMs, which are Mealy FSM and Moore state machine. In Mealy state machines, the output of the system depends on the current state of the system and the external inputs. If the output of the system depends on only the current state of the machine, it is called Moore FSM. Most of the FSMs found in SoC design are Mealy machines. Figure 4.4 shows Mealy FSM and Fig. 4.5 shows the Moore FSM.

In synchronous digital systems, the system state outputs are synchronous with the clock signal. Most FSM-based systems are synchronous and the design procedure of synchronous systems are standard and more mature. In synchronous processor systems, the operations like instructions, executions, logic, and storage functions operate in synchronism with the system clock. In communication systems, data transmission and reception happen in synchronism with a clock. Figure 4.6 shows the timing diagrams of a few such operations. These require resetting logic to start the design in a known default state. Resetting logic can be asynchronous or synchronous to the clock.

A SoC can have many of large functional cores each operating with a clock of its own as shown in Fig. 4.7.

The generation of the clock and its distribution to all the sequential elements of the SoC design have a significant impact on the performance and power dissipation of the SoC. For proper operation of the system, it is necessary that the clock edge arrives at all the clock inputs of design elements in the SoC design at the same time. But due to interconnect effects at submicron technologies, there will be a spatial shift at the clock edges when they arrive at the clock inputs of different timing elements. This results in a phase shift with reference to the source clock. This spatial shift in arrival time of the clock transition at different locations in the SoC (edge 1 in figure arriving at edge 2) is called clock skew as shown in Fig. 4.8. The clock period

Fig. 4.4 Mealy finite-state machine

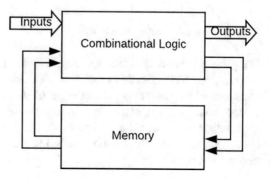

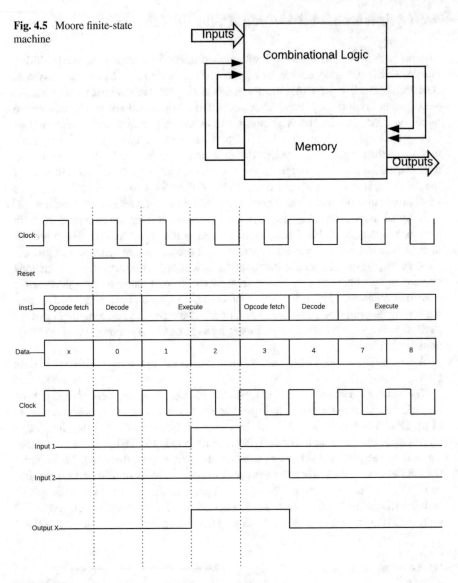

Fig. 4.5 Moore finite-state machine

Fig. 4.6 Timing diagrams of synchronous systems

in a given point of time at the same clock input of a design element in a chip can also vary in time. This is called clock jitter. The clock skew and clock jitter together constitute clock uncertainty. The design of the clock distribution network in SoC design should ensure that the clock skew is considered in meeting the setup and hold requirements of sequential elements in the design. Aside from timing closure addressing metastability is important during the VLSI logic design for synchronous sequential circuits.

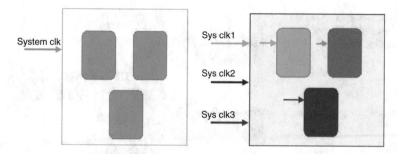

Fig. 4.7 Synchronous SoC blocks

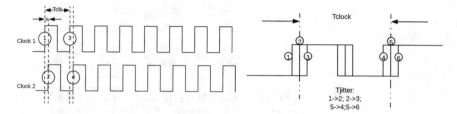

Fig. 4.8 Clock skew x and clock jitter

4.2   Asynchronous Circuits

System logic could also be designed as asynchronous circuits with reference to the system clock. These are made up of asynchronous combinational circuits and asynchronous signals. The output of the logic depends only on the inputs at the time, as against the synchronous logic, where the output of the logic changes with the inputs at the clock reference. Asynchronous logic outputs are difficult to predict in complex systems as they are traceable only to inputs, which can change at anytime. An adder, comparator, and a multiplexer/demultiplexer are a few examples of asynchronous logic circuits. Figure 4.9 shows the adder circuit and its timing diagram. Hence, debugging a SoC issue due to internal asynchronous logic circuits is difficult. Systems are realized with many smaller sets of combinational logic circuits, which are synchronized with clocks at appropriate levels to make them predictable and debuggable. These systems are called globally synchronous and locally asynchronous systems (GSLA systems).

4.3   Speed Matching

If multiple signals are crossing over domains of different frequencies, it is required to be double synchronized with the clock of the receiving domains to ensure that they do not become indeterministic. It is very common in large systems to have different subblocks operating at clocks of different frequencies and hence different

Fig. 4.9 Adder as asynchronous logic with its timing diagram

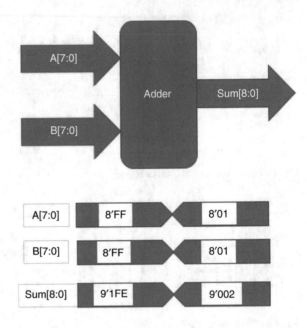

Fig. 4.10 Speed matching using FIFO

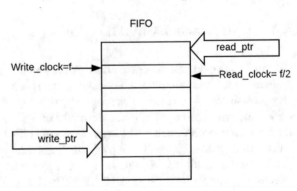

speeds. In this context, it is necessary to add speed matching logic if there are many related signals or multiple data lines crossing the domains of different speeds. The easiest speed matching technique is to use asynchronous first in first out (FIFO) as shown in Fig. 4.10 with source clock of a subblock writing and the destination clock of a subblock reading. The FIFO threshold is maintained safely to the extent of the clock speed difference. That means, by design, write access to the FIFO is permitted only if the previous data written is read out. The FIFO technique of speed matching is used in all communication protocol SoCs in cases where the transmit and receive clocks differ either in frequency or phase or both.

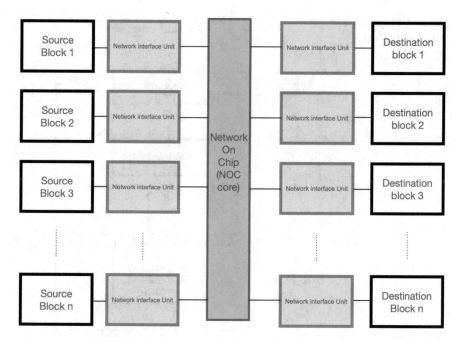

Fig. 4.11 SoC with NOC core

4.4 Network on Chip Architecture

For less complex ASIC designs, the hierarchical or crossbar bus architecture was conveniently used to get the desired performance. With the growing complexity of SoC, hundreds of IP cores get interconnected in a system. PPA requirements are never a compromise. Many SoCs integrate many numbers of protocols into the same chip which involves heterogeneous message transactions across the buses. The best way for such SoCs is to interface the large number of IPs or functional blocks using network on chip (NOC) cores. This circuit gets the transfer messages from one of the blocks to the other using network interfaces. It converts all bus transactions from source blocks to standard commands, which are decoded, the recipient block is captured, and transaction level data is transferred efficiently. A typical NOC core architecture for SoC design is shown in Fig. 4.11.

4.5 Hardware Accelerator

Certain functions in SoC do not require full implementation in hardware because they are not time critical. The parts of the functions that are time critical are implemented in hardware, and the partially processed data is accessed by the software to complete

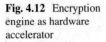

Fig. 4.12 Encryption engine as hardware accelerator

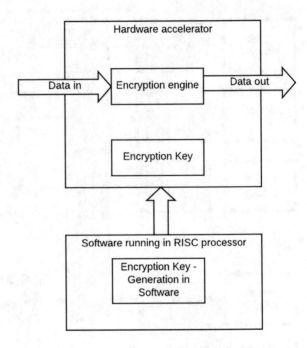

the function. The hardware section of the functional block which partially processes the data is called the hardware accelerator. The best example of the hardware accelerator is the encryption engine, which is the time critical part of security function that becomes the hardware accelerator. This encrypts the incoming data in real time using the key configured in the hardware. The key generation part is implemented in software running on general RISC processor core. Figure 4.12 depicts an encryption accelerator engine for security features in a SoC.

4.6   Hardware Description Languages (HDL)

Design methodology has evolved so much in the last six decades, so it has the complexity of SoC designs. A major part of the design evolution has to be attributed to the development of hardware description languages and EDA tool algorithms which can decode and process them to synthesize the equivalent logic by mapping it to the target standard cell library, making it fabricatable. To appreciate the modelling procedure using hardware description language, it is essential to understand the difference between hardware and software implementation. Table 4.1 lists the differences between hardware and software.

From the table, it is obvious to understand that the hardware description language should bear a minimum and support concurrent logic structures and has to have the concept of timing as against the software system description language

Table 4.1 Difference between hardware and software

Sl. no.	Hardware	Software
1	Concurrent execution of tasks. This demands all tasks and events to operate in coherence with a timing reference signal called clock	Sequential execution of tasks and instructions. There is no concept of synchronization to clock reference
2	Very fast execution. Functional timing in nanosecond scale units is achievable in hardware. And therefore, time critical functions are designed to be in hardware	Slow execution. Minimum timing resolution is 100 s of microsecond
3	Can be parallel	Sequential though it can appear to be parallel for the user
4	Physical and costs are exorbitant if it has to be redone	Can be recompiled
5	Need to be first time success	Can be corrected and recompiled without much effort
6	Hardware can be one time developed as platform and reused for lifetime if the functionality is the same	Can be redone easily
7.	Development from paper specification to physical system on chip	Need processing hardware platform for sw development
8	Need to verify fully imagining all scenario ahead of fabrication and hence verification, and validation is unavoidable	Verification is necessary to prove the intent of the design but in the case of minor defects, it can be corrected

called high-level programming language (HLL). This demands an understanding of the hardware realization to be able to model it using HDL. Most commonly used HDLs to model the SoC designs are SystemVerilog and VHDL. Language reference manual from IEEE standard association defines the requirement of HDL as language which should be "both machine readable and human readable, should support the development, verification, synthesis, and testing of hardware designs, the communication of hardware design data, and the maintenance, modification, and procurement of hardware." The reader is advised to go through hardware description language books given in reference to master the semantics and syntax of the constructs supported by the language as only relevant material is covered in this book. For the language reference manual, reader is encouraged to refer to IEEE documents from the IEEE standard association official site, describing the hardware design is termed RTL (register transfer level) design. This represents the functionality or design intent as a set of register transfers. This representation is most commonly used in the industry, which follows standard cell-based design methodology. The design flow is process technology (foundry) independent for getting the standard cell library from the foundry. Depending on the style of hardware description, models are classified as **behavioural modelling, dataflow** and **structural modelling**.

4.7   Behavioral Modelling of the Hardware System

If the functional behavior of the hardware is modeled using Verilog or VHDL, it is called a behavioral model. Examples of behavioral models of a simple decade counter and multiplexer in Verilog and VHDL are given in Fig. 4.13.

When the hardware is behaviorally modelled, it has to be synthesized to a gate level netlist. It is therefore necessary to make the model synthesizable. This is called the synthesizability of the model. Though coding for synthesis comes with experience, there are tools which check if the model is synthesizable. These tools are called lint tools. Using behavioural modeling, any complex functionality of the system can be represented, and by making it synthesizable, it can be transformed to gate level netlist. The structural description of the SoC netlist file is written using HDL constructs.

4.8   Dataflow Modeling of the Hardware System

A system can also be modelled as dataflow where the data progresses with different processing from different layers in a particular direction. These can be found out in communication systems. These models are also synthesizable. In the primitive sense, the dataflow model is the modelling sequence of the logic functions applied on the input data to arrive at the desired output data. For example, the dataflow modelling using Verilog for the circuit shown in Fig. 4.14a is given in Fig. 4.14b.

4.9   Structural Modeling of the Hardware System

Structural modelling is the style where the hardware modules are instantiated and are interconnected to realize the function. HDLs, Verilog, and VHDL support structural styles of modeling. It is easy to instantiate and integrate the analog IPs in hard macro representation and PADs in structural style into the SoC design. A netlist output by the synthesis process is the structural modeling of the hardware system using cell libraries, hard macros, and memory macros. Synthesis and physical design tools write out netlist in this style. An example of a structurally modelled code is shown in Fig. 4.15. SRDFF, INV, and ADD are the cells from the standard cell library. In this style, the standard cells are instantiated, and signals are interconnected to get the desired function. This is possible if the circuit design is of small complexity, or else it becomes extremely difficult to design the functional module in this style. However, in the case of a hard macro, the internal logic is protected, and only interface signals are provided which can be interfaced with the interfacing guidelines of the hard macro.

// Verilog model of decade counter

```verilog
module decade_counter (en, clock, rst,count);
input en, rst,clock;
output reg [3:0] count;

  always @( posedge clock or negedge rst )
    if (!rst)
      count <= 4'h0
    else
    begin
      if(en)
        begin
          if ( count >=4'h0 && count < 4'hA) //hex A = dec 10
          count<=count+4'd1;
            else
                count<=4'd0;
        end
      else
        count<=4'd0;
    end
  endmodule
```

// VHDL model of decade counter

```vhdl
library ieee;
use ieee.std_logic_1164.all;
use ieee.std_logic_arith.all;
entity decade_counter is
port (en,clk,rst : in std_logic;
count : out unsigned(3 downto 0));
end counter2;//port definition
architecture decade_counter_bhv of decade_counter is
signal reg : unsigned(3 downto 0);
begin
process(clk,rst)

begin
if rst='0' or reg="1010" then
reg <= "0000";
elsif (en ='1') then
```

Fig. 4.13 Behavioral model of decade counter in Verilog and VHDL

```
  if (clk'EVENT) and (clk='1') then
    reg <= reg + "0001";
    end if;
endif;
end process;
count <= reg;
end decade_counter_bhv;
```

```
//Verilog model of Multiplexer for shared logic
module sharing_example (a, b, c, d, cond, y);
parameter w = 16;
input [w-1:0] a, b, c, d;
input cond;
output [w*2-1:0] y;
wire [w*2-1:0] a_times_b = a * b;
wire [w*2-1:0] c_times_d = c * d;
assign y = cond ? a_times_b : c_times_d;
endmodule
```

```
//VHDL model of Multiplexer for shared logic

library ieee;
use ieee.std_logic_1164.all;
use ieee.numeric_std.all;
entity sharing_example is
generic (w : natural := 16);
port ( a, b, c, d : in unsigned (w-1 downto 0);
cond : in std_logic;
y : out unsigned (w*2-1 downto 0) );
end sharing_example;
architecture rtl of sharing_example is
signal a_times_b, c_times_d : unsigned (w*2-1 downto 0);
begin
a_times_b <= a * b;
c_times_d <= c * d;
y <= a_times_b when (cond = '1') else c_times_d;
end rtl;
```

Fig. 4.13 (continued)

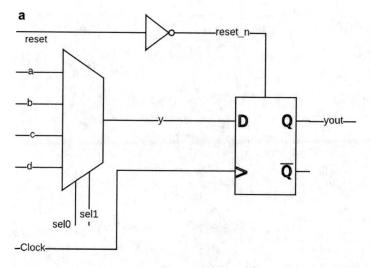

```
module mux4x1{clock, reset, a, b, c, d, sel0, sel1, yout};

input clock, reset, a, b, c, d, sel0, sel1;
output yout;

reg y;
wire reset_n
wire a, b, c, d, sel0, sel1;

  assign   y => a & ~sel0 & ~sel1;
  assign   y => b & sel0 & ~sel1;
  assign   y => c & ~sel0 & sel1;
  assign   y => d & sel0 & sel1;

  assign reset_n = ~reset;

always @ (posedge clock or negedge reset_n)
  begin
   if (~reset_n)
     yout => 1'b0;
   else
     yout => y;
  end

endmodule
```

Fig. 4.14 (a) Example circuit for dataflow modeling. (b) Dataflow modeling in Verilog for the circuit shown in Fig. 4.14a

```
module counter5(clk, reset, count, SRPG_PG_in);
 input clk, reset, SRPG_PG_in;
 output [4:0] count;
 wire clk, reset, SRPG_PG_in;
 wire [4:0] count;
 wire \count[0]_29 , \count[1]_30 , \count[2]_31 , n_0, n_1, n_3, n_4, n_5, n_6, n_7;

 SRDFF \count_reg[3] (.RN (n_3), .CK (clk), .D (n_7), .SI (n_1),
   .SE (count[3]), .RT (SRPG_PG_in), .Q (count[3]));
 SRDFF \count_reg[2] (.RN (n_3), .CK (clk), .D (n_6), .SI (1'b0),
   .SE (1'b0), .RT (SRPG_PG_in), .Q (\count[2]_31 ));
 ADD g103__8780(.A (\count[2]_31 ), .B (n_4), .CO (n_7), .S (n_6));
 SRDFF \count_reg[1] (.RN (n_3), .CK (clk), .D (n_5), .SI (1'b0),
   .SE (1'b0), .RT (SRPG_PG_in), .Q (\count[1]_30 ));
 ADD g105__4296(.A (\count[0]_29 ), .B (\count[1]_30 ), .CO (n_4),
   .S (n_5));
 SRDFF \count_reg[0] (.RN (n_3), .CK (clk), .D (n_0), .SI (1'b0),
   .SE (1'b0), .RT (SRPG_PG_in), .Q (\count[0]_29 ));
 INV g110(.A (\count[0]_29 ), .Y (n_0));
 INV g112(.A (n_7), .Y (n_1));
 INV g114(.A (reset), .Y (n_3));

 endmodule
```

Fig. 4.15 Structural modeling style

4.10   Input-Output Pad Instantiation

As shown in Fig 4.16, structural code of Input-output pads for signal and power are instantiated from the pad cells in the target library. Standard practice is to add them in the top module of SoC hierarchy.

4.10.1   Power Ground Corner Pad Instantiation (Fig. 4.17)

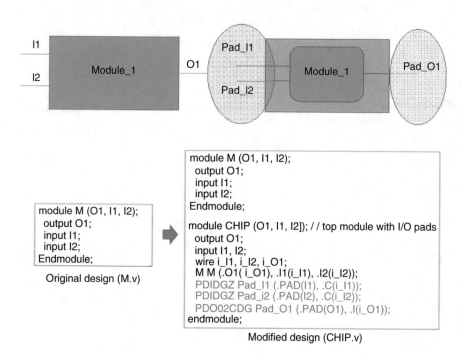

Fig. 4.16 IO pad integration

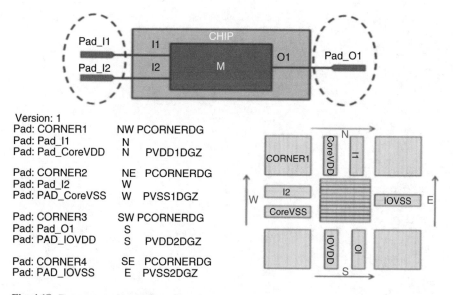

Fig. 4.17 Power ground pad integration

Chapter 5
Synthesis and Static Timing Analysis (STA)

5.1 Part 1: SoC Synthesis

The process of converting a behavioural RTL model of a system to a structural (logical netlist) model using synthesis tool is called synthesis. The synthesis covered here is with respect to the automatic cell-based design process. The conversion is done in two steps: in the first step, the behavioural representation of the design is converted to generic netlist, and in the second step, generic netlist is converted to the netlist using cells from the target standard cell library, also called technology library. Standard cell library contains all the design files of a set of standard cells (universal logic gates or primitive modules), which are pre-designed, verified, and characterized by the foundry. The library includes behavioural models, timing models, and physical models of the standard cells. They are targeted at a particular manufacturing process used in the fabrication by the foundries. The fabrication houses design, validate, manufacture, and process devices on silicon wafers using SoC designs done based on an automated cell-based design methodology. The foundries provide technology cell libraries for performing SoC designs. The standard cell characteristics in the technology library data correlate with the standard cells processed on the base wafers. The design files from the cell library are used during design synthesis, verification, timing analysis, physical design, physical design verification, power, and parametric analysis of the designs. Similarly, the input-output (IO) pads are also characterized for electrical and physical parameters and are available as IO or pad libraries. The standard cell library and pad library are reused for multiple SoC designs targeted at the same process technology. The synthesis tools use advanced high-tech conversion and optimization algorithms to map the behavioural RTL design to design netlists using cells from the standard cell library. The SoC design netlist during the synthesis process is optimized by removing redundant logic and logic sharing in the design, retaining the design intent. Optimization is carried out using a synthesis tool which uses advanced logic optimization algorithms. Figure 5.1

© The Author(s), under exclusive license to Springer Nature Switzerland AG 2022
V. S. Chakravarthi, *A Practical Approach to VLSI System on Chip (SoC) Design*,
https://doi.org/10.1007/978-3-031-18363-8_5

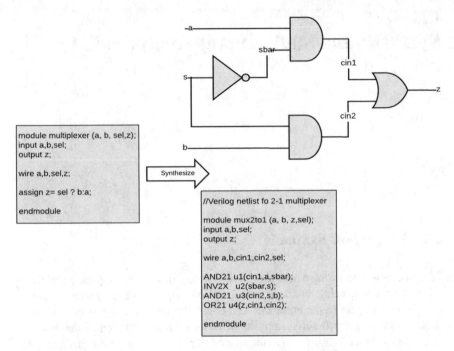

Fig. 5.1   Process of synthesis

depicts the process of synthesis. As it can be seen, the gate level netlist generated by the synthesis tool is the structural representation of the behavioural description of the SoC design. Hence, it is necessary that the behavioural model of SoC design in RTL code is synthesizable. This demands correct use of HDL constructs for the functions in RTL code. Only a subset of the HDL constructs is synthesizable, and care should be taken to ensure that the RTL code is synthesizable. This is typically verified by the lint tools. The process is called linting. It uses a set of defined rules to check the RTL module for synthesizability, testability, and redundancy.

Logic synthesis flow using the synthesis tool is shown in Fig. 5.2. The design inputs for the synthesis process are RTL design files, the standard cell technology library, and design constraint file. The RTL representation of the SoC design is a set of RTL files corresponding to system modules, functional and memory macros, and IP cores. The standard cell library is a library of pre-validated logic cells, with the circuit parameters and process parameters. Some of the standard cells are a set of logic cells like INV, buffer, AND, NAND, OR, NOR, XOR, XNOR D flip-flops, latch cells, complex cells like multiplexer (MUX) cells, encoders, decoder cells, and complex cells like adders, multipliers, etc. Design constraints are design clock definitions, input-output signal delays, and user preferences, for the synthesis tool to generates netlist for the specific PPA goals. PPA goals for the design are timing, area, and power requirements. Synthesis is done in two stages. In the first step, it maps the RTL design to generic cells from the GTECH library. GTECH library is a

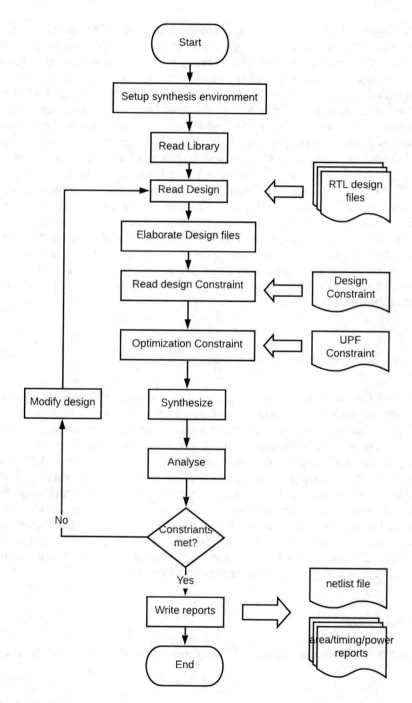

Fig. 5.2 SoC Synthesis flow

virtual logic library with generic logic cells with defined logic functionality. The synthesis tool performs first-level logic optimization on the GTECH-based design netlist using digital logic optimization techniques. The optimized GTECH netlist is then mapped to the standard cells from the technology library in the second step. Synthesis optimizes the mapped netlist at this stage to achieve set design goals. The optimization algorithms use two-level and multi-level optimization techniques, combination of sequential synthesis, and logic sharing techniques. More on the theory of synthesis and optimization algorithms, user can refer to synthesis and optimization of digital circuits by Giovanni De Micheli, Tata McGraw-Hill edition. Synthesis flow also accepts power intent of the design using a *unified power format* (UPF) for power optimization. This requires special types of power management library cells in the standard cell library. Some of the special power management cells are level shifter cells, isolation cells, always-on cells, etc. Different steps in the synthesis flow are explained below. Figure 5.2 shows the SoC level synthesis design flow. The synthesis tool writes out the optimized design netlist, updated design constraint, and design area, timing, and power reports. Designers at this stage will do the first level of design check for logic equivalence and report analysis for design performance against set goals.

5.1.1 Set Synthesis Environment

The synthesis environment is set up the directory structure for the synthesis process by defining paths of RTL design files, design constraint files for timing, area and power, and the choice of standard cells to be used. The RTL model design files are in hardware description language (HDL) format, design constraint file is in standard delay format (SDF), and a power constraint file is in unified power format (UPF) format. The SDC file contains clock definitions, input-output delays, maximum fanout capacitances, and multiple cycle paths and false paths in the design, etc., and the UPF file consists of design partitions like always on blocks, power domains, and switchable and non-switchable power supplies in the design. Synthesis setup also defines the name and the directory path where the output design netlist, report files, and synthesis logs are to be saved. Typically, the library conversion into a specific format as accepted by the synthesizer is done in this stage.

Read Library

Synthesis tool reads the standard cell library, macro library, and the IO cell libraries.

Read HDL Design Files

The behavioural models of the SoC design as RTL files in HDL format are read into the synthesis tool. The most commonly used HDLs for SoC design are Verilog, SystemVerilog, and VHDL. The design files have the file extensions of v for Verilog, .sv for SystemVerilog, and .vhd for VHDL.

Elaborate the Design Files

This phase elaborates the design such that multiple usages of the functional modules are uniquely resolved. This stage identifies logic functions and ensures that every other module can access the functions when they need them without any conflicts. The tool does optimization by sharing common logic, removing redundant logic, identifying registers, designing elements in the target library, etc. Design parameters are identified, resolved, and finalised at this stage.

Read Design Constraints

The design constraint file has definitions of clock details and other signals like clock source and input-output delay specifications. The file contains clock frequency, uncertainty, and its relationships to the rest of the generated clocks used in the design subbocks. Specific input-output delays, false paths (redundant paths), and multicycle paths are listed and read in. Some of the design details, such as clock domain definitions and their inter-relations etc., such that the tool can consider during design optimization. The design realization uses suitable standard cells is guided by the design constraints. The design constraint is input to the synthesis tool in the standard delay constraint (SDC) file format. In brief, it consists of the design clock definition, grouping of the clocks (clock domains), and applying design rule constraints (DRC) like maximum transition times for the signals. An example SDC file are shown in Fig. 5.3. In the constraint below, the text after # is the comment of the constraint statements.

```
current_design top
# module design hierarchy for synthesis is set
set_units -time 1000.0ps
# sets time resolution
set_units -capacitance 1000.0fF
# sets load resolution
set_clock_gating_check -setup 0.0
# setup constraint for clock buffer
create_clock -name "clk" -add -period 7.0 -waveform {0.0 3.5} [get_ports clk]
# clock signal generation with period 7ns and pulse width 3.5ns (50% duty
cycle) to apply at design port clk
set_input_delay -clock [get_clocks clk] -add_delay 0.3 [get_ports clk]
# clock signal input delay constraint to account for clock uncertainty.
```

Fig. 5.3 Extract of the design constraint in SDC format

Optimization Constraint

The primary SoC design goals like power, timing, or area, generally referred to as PPA (power, speed performance, and area), are specified and fed in as the desired optimization setting for synthesis. Based on the configured design goal, the design is implemented. The selection of cells used for logic conversion depends on the design goal setting during synthesis. The synthesizer maps the design logic to the design elements in the standard cell library, meeting the constraints. Needless to mention, the standard cell library will have many choices of design elements and types of standard cells to map to realise the function. Low-power SoCs have become the utmost necessity today. The optimising power and areas were the design goals of the past in higher technology nodes as they were achieved by design techniques. These in current deep sub-micron technologies are guaranteed by default. But achieving low power for the high density systems on chips has been the new challenge. This require system level power optimisation techniques to be applied during the design., The power strategy at the SoC level is written as power constraint for the design. The power intent of the design is fed into the synthesizer in unified power format (UPF) file. UPF is a standard format defined by Accellera which is published as an IEEE standard. The most recent version of the standard is IEEE1801-2013. UPF for SoC design allows definition of power management strategies with multiple power supplies in SoC design. You can define different power domains in SoC design, all the necessary resources for the signals to cross the power domains, the insertion of level shifter cells or isolation cells across the power domains, and switches to turn on and off the power supplies for managing power in the design.

Synthesis

After all the necessary input design files are read into the synthesizer tool, the RTL design is converted to a generic netlist using GTECH library cells and then mapped to a netlist using standard cells from a target cell library. The GTECH library contains a set of technology-independent general logic cells. The first level of logic optimization is performed at this stage.

Analyze

The generated design netlist is assessed against the design constraints, specified optimizations, and design errors and warnings if any. Any ambiguous RTL logic that the tool cannot convert to an equivalent netlist will be written out as errors or warnings.

Generate Reports

The design netlist, area, timing reports, and the updated design constraint files are written out in files in the folder identified in the environment setup. The synthesis errors and warnings are to be analyzed and the design fixes are provided. The updated design is reworked. It is essential to check if there are any errors in design conversion. Also, there will be a large number of warnings reported after analyzing the design conversion, which need to be fixed. This is easily done using scripting languages. Working knowledge of scripting languages like Perl and tool command language-tool kit (TCL-TK) helps in analyzing large size log files the synthesizer writes out.

There are two kinds of activities in the Synthesis step. The first one being design conversion and report generation. The synthesis tool supports different commands with many options for these activities.

5.1.2   SoC Design Constraints

The SoC design is synthesized with a specific design constraint to make it operate at the specified range of operating frequencies (timing constraint) or restrict it to a size (area constraint) or use a set of standard cells or combination of them to achieve a low-power design (unified power constraint). The synthesis tool accepts the constraint file along with the design files to generate optimized design blocks. The timing file extracted from the design during synthesis is also fed to the timing analysis and simulation tools for design verification with back annotation.

Consider the design example shown in Fig. 5.4.

The interface signals are shown in Table 5.1.

An indicative design constraint file for synthesizing above design is shown in Fig. 5.5.

The design constraint file contains the clock definition and its parameters such as the clock latency, the clock uncertainty, and input-output delays of other signals in the design. The sample of the design constraint file is as shown in Fig. 5.5. The

Fig. 5.4 Example design for the synthesis

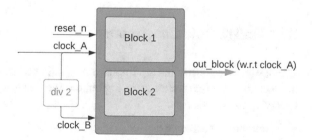

Table 5.1 IO signal description table for the design in Fig. 5.4

Signal name	Input/ output	Bit width	Description	Default
clock_A	Input	1	Master clock of frequency 50 MHz	1'b0
reset_n	Input	1	Active low reset; design will be reset to the default values when this signal goes low	1'b1
clock_B	Input	1	Derived clock from clock A. it is the divide by 2 of the frequency of clock A of 50 Mhz. Its frequency is 25 Mhz	1'b0
out_block	Output	3	Timer output w.r.t. clock_A.	3'd0

```
#creating master clock for clock_A for 50MHz
define_clock -domain domain1 -name clock_A -period 10 [find / -port clock_A]
#creating generated clock for clk_B from clock A to get 25Mhz frequency
define_generated_clock [-name clock_B]-source [get_port clock_A][-divide_by 2] –get_ports clock_B
# setting ideal network attribute
set_attribute ideal_network true
get_attribute ideal /B1/clock_A
#setting input latency for the clock to model the pin delay or the delay due to the CTS
set_clock_latency 2m [get_clocks clock_A]
#setting the clock skew because of Clock tree synthesis
set_clock_uncertainty -setup -rise -fall 2m [get_clocks clock_A]
# Constraining outputs with the delay due to the pin propagation delay
set_output_delay 4 [out_block]
```

Fig. 5.5 Design constraint in SDC file format

constraint file also contains the maximum limit on fanout and net capacitance, any restrictions on the standard cells to be used to map the design. The typical restrictions on cell usage for Synthesis include avoiding cells with low drive strengths, permit the use of the standard cells from two or more standard cell libraries etc.

Synthesis tool generates the most optimized design netlist from RTL design files considering the design constraint and technology cell library files. To a great extent, the tool tries to optimize the netlist such that the no setup and hold violations are found on the timing paths and don't exceed given area and power targets as defined in the design constraint file. However, there can still be violations on all set goals which require designer's intervention to fix. The present-day synthesis tool uses analysis tools such as timing analyzers and design rule checkers for design mapping and optimization.

5.2   Design Rule Constraints

The design rule constraints are imposed on synthesis by the physical limitations of the technology library chosen to implement the design. Design rules include the following three elements:

• Maximum capacitance per net.
• Maximum fanout per gate.
• Maximum transition time of the signal.

These three constraints are used together to ensure that the library limits are not exceeded in the design. A good designer studies the library property of the cell library and constraints of the design so that the design meets the design goals with the least number of iterations.

5.3   SoC Design Synthesis

Behavioural synthesis is also called architectural synthesis or high-level synthesis. It involves identifying architectural resources needed for the behavioural representation of the SoC design, and binding the resources to the functions, and determining the execution sequence or order of execution. To achieve high-quality (performance) netlist representation, the synthesis activity should be strategized keeping in mind the following:

- Complexity of the SoC.
- Number of subsystem blocks in the SoC.
- Number and types of IP cores: soft, hard, and netlist.
- Capability of the computational system on which the synthesis is run.
- Debug capability of the designer.

Design synthesis writes the netlist with updated constraints and design reports so that the designer can verify the netlist to retain the design intent. It also writes out the audit log of the set of processes it has done with the appropriate errors and warnings where it has violated and not met the design constraint. When the SoC complexity is high, it is good practice to synthesize the design with two or three levels of hierarchy. This helps to retain the module names in the design netlist and eases the debugging of the logic nonequivalences, if any. The synthesizer tool can write out the netlist either in hierarchical, with the level of hierarchy maintained in the input file, or flat netlist where the entire design hierarchy is collapsed into a single level. If the SoC design is of low complexity, it is synthesized in one execution with all the modules at the same level of hierarchy. This is called flat synthesis. The entire design will be converted to a gate level netlist with the same level of hierarchy as the smallest standard cell. Though the netlist is in a readable HDL format, it will be in a flattened hierarchy with the set of instances and interconnects. All the instances and interconnected net names are tool generated. This makes it difficult to identify the logic functionality and correlate it with the RTL design. Debugging such a flat design netlist is very difficult and time-consuming.

In hierarchical synthesis, design at block or module level, as per the hierarchy maintained by the designer, is synthesized one by one, and then all the block level netlists are read into the tool along with just the top-level module and written out as the hierarchical or flat netlist as required. Any core available as a netlist is read into the tool and the final netlist is updated. Hard cores, if read into the tool, will be a black box with only interface connections and without any functionality. It is therefore necessary for the designer to have knowledge of all the SoC instances. Along with the netlist, the synthesis SDC is also written out which is to be fed in along with the netlist to static timing analysis (STA) tool and physical design tools.

It is during the synthesis that all the flip-flops of the design are replaced with scan flip-flops from the library to enable DFT activity (which will be discussed in the next chapter). To optimize the SoC design netlist, it is essential to direct the tool through the SDC constraint file to use a certain set of standard cells (restrict it from using some low-drive standard cells) and mixed set of high-performance logic cells from the same library depending on the design goals. An example of this is the use of low and high threshold voltage (Vt) cells in appropriate modules to get a low-power netlist.

5.4   Low-Power Synthesis

Design is synthesized for low power as a design goal, which requires additional design constraints in universal power format (UPF). UPF defines the power management strategies by identifying always-on block, switchable blocks, and conditions at which the switchable blocks are controlled and reliably without breaking the functionality.

5.4.1   Introduction to Low-Power SoCs

Power consumption of systems has become one of the most important figures of merits of the SoC designs. SoC power management has become a major requirement for SoC design as power density has grown to alarming figures, raising questions about the feasibility of design implementation. It is possible only if power management requirements are considered at every stage of SoC design, right from the architecture definition stage to the design tape out. The power density trend versus power design requirements for modern SoCs is mapped in Fig. 5.6. The widening gap represents the most critical challenge that SoC designers face today.

In some of the nanometer technology cell libraries, the cell leakage power is greater than the switching power, demanding an aggressive power management strategy for SoC designs. Operand isolation, clock gating, multi-VT designs, multiple supply voltage (MSV) designs, dynamic voltage and frequency scaling (DVFS), and optimization of clock tree synthesis (CTS) is one of a few techniques of power management in the SoC. In-depth treatment of power management is not the scope of this book. However, to achieve low-power SoC design, it is essential to define the power intent of the design in addition to the design intent and define by design at all stages of design, including synthesis. The low-power SoC design flow involves defining the correct power intent and successive refinement method as design advances, as shown in Fig. 5.7. Power constraints in UPF define the power

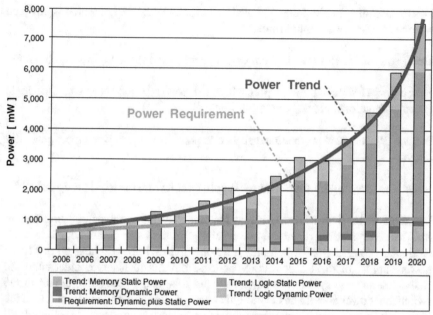

Fig. 5.6 IC power trends: actual vs specified. (Courtesy: Si2 LPC)

```
#---------------------------------------------------------------------
#              Create power domains
#---------------------------------------------------------------------
#Connect top level ports with supply sets defined in power domains created
#---------------------------------------------------------------------
#              Define required power switches with switch conditions
#---------------------------------------------------------------------
#              Set isolation strategies
#---------------------------------------------------------------------
#.             Define isolation details and rules
#---------------------------------------------------------------------
#              Set retention strategies
#---------------------------------------------------------------------
#              Set level shifter strategies
#---------------------------------------------------------------------
```

Fig. 5.7 UPF content for power intent specification

distribution management, design partitioning into regions using independent power supplies, and interfaces and interactions between these regions. To understand the process of defining the power intent in UPF format, it is necessary to understand a few terminologies used in the power context. A few important ones are defined in this section.

Power Domain Logic group or functional blocks in the design which is powered from the same power supply source.

Drivers Ports or nets on rail, from which power is fed to the logic group or block.

Receivers The receive net or port where the power is first received in the logic group or functional block.

Source Power source is the first distribution point from the power supply generator circuit.

Sink The receive path for the power supply circuit from the logic group or block.

Isolation Cells Power management for achieving low power consumption in SoC involves shutting down the power supply of the logic circuits operating in a particular power domain when they are not active. While switching on or off the power supply, there is a possibility of logic circuits to get into indeterministic states, making the system unstable. To avoid this danger, the logic circuit in the switchable power domain has to be held at known state and isolated from rest of the logic circuits, and then, the power supply must be shut down. A special standard cell in the library which isolates the power domain is called an isolation cell. It should be ensured that logic is safely brought to known state and then power is switched off. The set up for making the power domain switchable is shown in Fig. 5.8.

Level Shifters In SoC design, different power domains operate at different voltages driven by different power sources, and the signals crossing the domains are to be set to appropriate power levels in their respective power domains. This is accomplished by level shifters. These are special cells in the cell library which can boost the power or buck the power to the appropriate level as required in the SoC design.

State Retention Before the logic blocks or functional cells are switched off when not in use, the design will retain some of the states of SoC, and restored when power is switched on. This is done by using the special cells called state retention power gating (SRPG) cells in the library.

Multi-VT Cells Power optimization is achieved using mix of multi-Vt cells in the design-netlist. They are cells of different threshold voltages in the design. The standard library for low power SoC designs has cells of different threshold voltage levels. Synthesis tool algorithm depending on the design need, use cells of appropriate threshold voltages. Low-Vt cells are applied for high-speed and high-Vt cells are mapped for noncritical paths. This is possible by using multiple libraries containing multi-Vt cells.

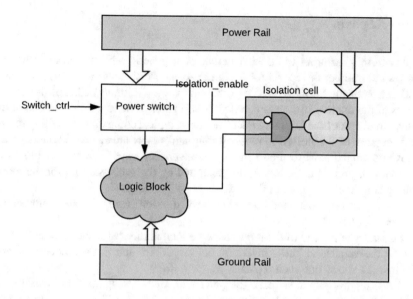

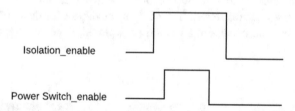

Fig. 5.8 Isolation cell and power switch for low-power SoC designs

5.4.2   Universal Power Format (UPF)

UPF file contains the power intent of the SoC design. In this file, we define the power regions with corresponding power supplies, always on block with a default power supply, interfaces and signal interactions across domains, and power management strategies like the requirement of state retention. The synthesis tool reads the UPF file along with the RTL and SDC files and generates the power-aware design netlist which includes appropriate level shifter cells, isolation cells, and power switches. Tools can also write out the modified UPF file, which can be used in further stages of design like P&R for power-aware physical design and analysis. The UPF file is used in logic equivalence check (LEC) for power-aware logic equivalence checks. A typical UPF file defines the following functions using appropriate commands which the synthesis tool can read.

5.5 Reports

Apart from generating the design netlist both generic and mapped, synthesizer writes out number of reports for design analysis. Most important reports are area and timing of the design. These reports will give a preliminary idea of the area in terms of the number of standard cell (NAND) gates or instances or in terms of the silicon real estate area in square micrometres. A typical command for writing timing and the area of the design is report timing and report area/gates. Variants of the above command exist to report these parameters for specific instances, blocks or subblock or path. The timing report generated by the synthesis tool for the report timing command is shown in Fig. 5.9.

The area report generated by the synthesis tool for the report gates command is shown in Fig. 5.10.

Summary at the end of the report shows a total number of instances and the area for all the sequential cells, inverters, buffers, logic, and timing models, if any. Figure 5.11 shows one such report.

These reports help to estimate the gate count, area. Timing margins in design can be used to further optimize based on the design goal chosen. The area report lists the total design area as well as a breakdown of the area per level of hierarchy. The area numbers are calculated by counting the number of cells instantiated in the design and multiplying by the cell area specified in the respective technology library. Refer to Fig. 5.11 for synthesis area report. If there is any deviation, the design files are to be modified to meet the constraint specified or explored if the constraints can be relaxed.

```
...
I1/clock
cout_reg_3/CK      <<<                                    0                0     R
          cout_reg_3/Q          DFFRHQX1   3   24.8  646   +518    518    R
I1/cout[3]
p0160A/B                                                  +0       518
p0160A/Y              NOR2X1     1   7.4    262  +174     692    F
p0201A/B                                                  +0       692
p0201A/Y              NAND3BX1   1   8.0    285  +174     866    R
p0257A/B                                                  +0       866
p0257A/Y              NOR4X1     1   3.6    185  +133     999    F
top_counter/flag5  <<<          out port                  +0       999    F
...
```

Fig. 5.9 Sample timing report showing timing of one of the design paths

```
=========================================================
Generated by: Tool:Version
Generated on: date
Module: top_counter
Technology library: GPDK slow 1.0
Operating conditions: slow
Wireload mode: ZeroLoad
=========================================================
```

Gate	Instances	Area	Library
AND2X2	10	166.3	slow
AOI21X1	2	33.3	slow
AOI2BB2X1	2	46.6	slow
DFFRHQX1	13	910.7	slow
DFFRHQX2	3	260.1	slow
INVX1	2	20.0	slow
INVX3	2	20.0	slow
NAND2X1	3	29.9	slow
NAND3BX1	2	39.9	slow
NAND4BX1	1	23.3	slow
NOR2X1	4	39.9	slow
NOR3X1	1	16.6	slow
NOR4X1	1	20.0	slow
OAI2BB2X1	8	186.3	slow
XNOR2X1	2	59.9	slow

| total | 56 | 1872.7 | |
Type	Instances	Area	Area %
sequential	16	1170.8	62.5
inverter	4	39.9	2.1
logic	36	662.0	35.3

Fig. 5.10 Area report of the design module output by the synthesis tool. (Courtesy: Cadence for Genus tool)

5.5.1 Gate Level Netlist Verification

The gate level netlist verification is done by a thorough review of errors and warnings in synthesis and timing reports and fix them. It is essential to scrutinise the optimization logs reported during the synthesis run to ensure that no required logic is optimized or removed. Running gate level simulation for the verification scenario is impossible as it is very time-consuming. The netlist elements will have timing requirements for input-output and other design elements. Modeling these timing

report area command on tool writes out the area report as below:
===
Generated by: Tool and *version*
Generated on: *date*
Module: Module Name
Technology library: GPDK slow
Operating conditions: slow
Wireload mode: Zero Load
===

Block	Cells	Cell Area	Net Area	Wireload
Module Top	56	1873	0	Zero_Conservative
I2	24	880	0	Zero_Conservative
I1	24	863	0	Zero_Conservative

Fig. 5.11   Area report depicting the number of the instances

delays and clock abnormalities and cell delays and understanding the timing needs dynamically in simulation scenario is practically not possible. So, only a selected number of critical test scenarios are simulated with the design netlist considering their timing. Another most important check done to ensure correct generation of design netlist is logic equivalence check (LEC). Every time synthesis is done, it is essential to run the logic equivalence between the gate level netlist file generated by synthesis process and the golden reference RTL file, which is used as input to synthesis. The logic equivalence is verified using the formal tools. LEC tools also have a good debug facility to fix nonequivalences if any.

5.6   Part 2: Static Timing Analysis (STA)

Timing analysis is an important step in the SoC design process, which in a way differentiates it from software system development. In synchronous SoC designs, clock abnormalities (clock skew and jitter), interconnect delays, and timing requirements of sequential cells make timing analysis a critical step for correct operation. Analyzing design timing dynamically in different use case scenarios is practically impossible. Hence, static timing analysis (STA) is performed on all the design paths. This does not require input stimulus. The following are a few of the definitions and concepts required to understand static timing analysis (STA):

5.7 Timing Definition

Clock Signal Most of the SoCs are synchronous and operate in synchronism to the timing reference called clock. The clock signal is a periodic, repetitive waveform with a fixed frequency that will be used by the logic in the SoC design to time and sequence its operations. In SoC design, the clock is used as a reference signal to get events, states, and processed signal/data captured and propagated to the subsequent logic elements.

Design Objects Design objects are the logic blocks with input ports, output ports, and functional blocks realized using sequential cells such as flip-flops, latches, and combinational circuits.

Clock Latency Clock latency is the time delay seen between the clock edges of the source and the destination. This is also called network delay from the clock output from the source generating it to the point under consideration. This includes clock skew and jitter. It is modelled as an insertion delay seen on the clock. This is caused by mismatches, imperfections, process variations in driver cells in the clock distribution network, interconnect effects (cross talk) in submicron technology, and variations in operating conditions (variations in temperature and power supply voltage) and due to varying loads. Figure 5.12 shows the sources of clock abnormalities.

Clock Domain A clock domain is a group of logic circuits operating on a single clock or derived clocks that are synchronous to each other, allowing timing analysis to be performed between them. Timing between two clock domains will be considered asynchronous and no timing check will be performed across the clock domains; However, signals crossing the domains have to be carefully designed so that data transfers reliably across clock domains in multi-clock domain SoC.

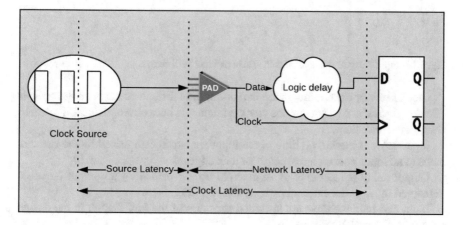

Fig. 5.12 Clock abnormalities

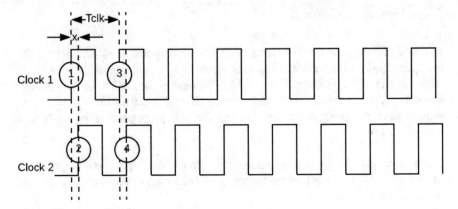

Fig. 5.13 Clock skew = x

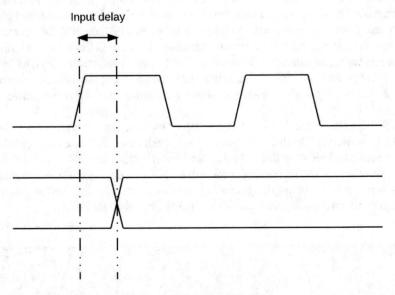

Fig. 5.14 Input delay of the input signals due to external path delays

Clock skew or uncertainty is the maximum time difference between the arrivals of clock signals at registers in one clock domain and between domains. Figure 5.13 shows the clock skew.

Input delay is the arrival time of the input signals is determined by the external paths at an input port with respect to a clock edge, as shown in Fig. 5.14.

Output delay is the delay of an external timing path from an output port to a registered input in the external path, as shown in Fig. 5.15.

Input and output delays are specified for ports of the SoC design in the design constraint file in SDC format.

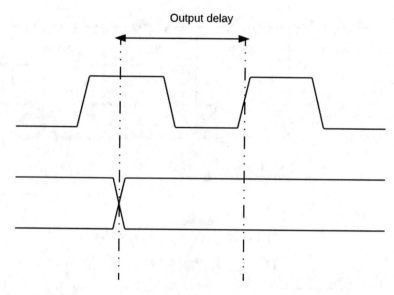

Fig. 5.15 Output delay associated with SoC outputs till they get registered externally

Fanout on Nets Limit on maximum fanout of any net can be assigned, which will be ten. That means that any net found in the design can drive a load equivalent to ten input cells. This will be used to map the right standard cell with the correct drive strength to the logic with the stated fanout.

Operating conditions like process, temperature, and voltage define the process variations, which affect the functionality and performance of the SoC design. For example, the higher the supply voltage, the smaller the delay; higher the temperature, and higher the delay.

Interconnect model is the parasitic parameter of the interconnect network for different sets of inputs and operating conditions, which are used to estimate the propagation delay of the path. There are many ways to represent an interconnect as a model and most common one to represent it as distributed resistance and capacitance as shown in Fig. 5.16. For analysis, wire segment with five to ten delay elements/nodes are considered for extracting the parasitics and path timing analysis. This is called the wire load model. Timing analysis is carried out considering the device propagation delay for the load connected to it.

Zero wire-load model represents zero net delays and is the pre-layout timing information of the design which shows only the propagation delays of the standard cells without the interconnect or wire delays.

A wire-load model is the net resistance and capacitance (RC) model used for timing analysis, and it provides an estimate of the RC load of nets computed for fanouts. Wire-load models are used to estimate the loading effect on interconnect delays in the design. By default, in an area-based wire-load model, the timing information is extracted from the technology library which will be used for timing analysis.

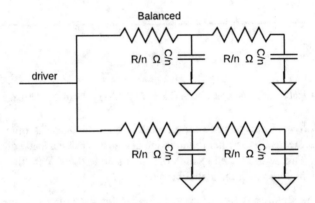

Fig. 5.16 Wire-load model for estimating resistance, capacitance, and pin capacitance

A false path is a path that will never be used during the operation of the SoC and hence it does not need to meet timing requirements. For the example shown in Fig. 5.17, if the select signals of the MUX1 and MUX2 are tied together, it is not possible for the valid path from input 1 of MUX1 to input 2 of MUX2. This path is the false path by design.

Architecturally, functional modes of SoC can have false paths across modes as no two modes coexist functionally in SoC operation. Signals that activate test modes are examples of false paths in the functional mode. Avoid timing violations by setting false path exceptions.

A multicycle path is a timing path that does not propagate a signal in one cycle. And in SoC design, it is not necessary that all paths have to meet single clock constraint, meaning the data launched with the launch clock edge need not reach the destination flop (capture clock) in a single cycle. For example, all the enables generated by the configuration registers will typically stay stable for multiple clocks, as shown in Fig. 5.18. They need not be closed for single clock. By default, static timing analyzer considers all paths to be single-cycle paths, and it is explicitly required to identify and notify to the tool the paths as multicycle paths in the design.

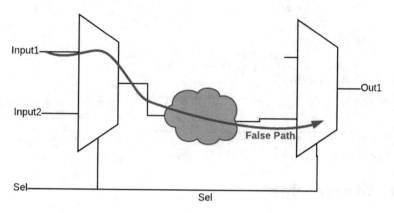

Fig. 5.17 False path example

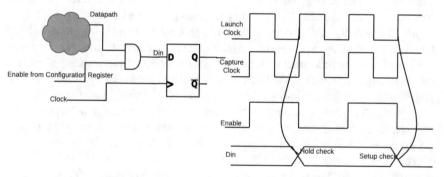

Fig. 5.18 Multicycle path example

SoC Functional Mode The functional mode of the SoC is the mode in which the SoC is designed to work independently as intended. There can be one or multifunctional modes for the SoC. Examples of multiple modes of SoC are low-power mode, fully functional mode, test mode, etc. In each of the modes, the frequency of the clock and timing requirements are different. It is required to analyze and fix timing violations, in each of the modes independently.

5.8   Timing Delay Calculation Concepts

The timing information of the cells and the net in which they are connected to neighbouring cells is listed in the library file in the form of a timing library format or TLF file. The reader can refer to the timing library format and the ways to analyze the path and cell delay from the standard TLF reference defined by Cadence. It defines the procedures for defining the timing model of a standard cell and computing the path delays, signal input, output slews, etc. Timing checks are the functions of cell delays and signal slews. A few timing parameters are shown in Fig. 5.19.

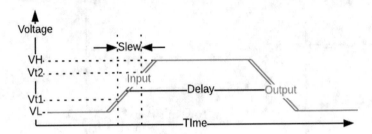

Fig. 5.19 Timing parameters

5.9   Timing Analysis

Timing checks can be done in two ways: dynamic timing analysis and static timing analysis. Dynamic timing analysis is the process of analyzing the SoC with actual functional vectors applied. This is a very cumbersome process, and also, it is highly impractical to apply all functional vectors and go through their timing along with the functionality. Also, dynamic timing analysis is next to impossible to assess at the gate level and for all the functional vectors. Static timing analysis is the process of analyzing the timing requirements of the independent paths without applying functional vectors. The SoC design is considered as a large set of directional paths from input to outputs, inputs to sequential elements like registers, and register to output signal paths, and then for each of the paths the timing requirements are analyzed using library timing details specified in the timing library format (TLF) file of the library cells.

The timing library format (TLF) file contains timing models and timing data to calculate I/O path delays, timing limits, and interconnect delays. Input and output path delay values for the library cells. The timing check values are computed on a per-instance basis and are called "cell-based delay calculation." for the design. Path delays in a circuit depend upon the electrical behaviour of interconnects between cells. The parasitic information in the TLF file is the estimated interconnect parasitics used for delay estimations of the design layout in the pre-layout stage of the design. Because actual variations of the operating conditions cannot be anticipated during characterization of delay data, derating models are used to approximate the timing behaviour of a particular cell at selected operating conditions. This includes process, voltage, and temperature models of the library cells at different operating conditions called grid points as in TLF data which are used to derate process, voltage, and temperature for off-grid points using interpolation or extrapolation equations.

In standard sequential cells like flip-flops, input signals need to meet certain requirements or limits for the physical cell to operate correctly. These limits, which are often functions of design-dependent parameters, like input slew or output load are used during the simulation to verify the operation of the cell. Models similar in concept to the delay or slew models are used to provide the data for computing timing checks.

Setup time: The setup timing check specifies acceptable range for a setup time. In a flip-flop, the setup time is the time during which a data signal must remain stable before the clock edge. Any change to the data signal within this interval results in a timing violation. Figure 5.20a shows a positive setup time—one occurring before the active edge of the clock and the difference between a positive and a negative setup time.

Hold time: The hold time check specifies limit values for a hold time. In a flip-flop, the hold time is the time during which a data signal must remain stable after the clock edge. Any change to the data signal within this interval results in a timing violation. Figure 5.20b shows a positive hold time and Fig. 5.20c shows positive setup and negative hold time scenario.

Skew The skew timing check specifies the limit of the maximum allowable delay between two signals, which if exceeded causes devices to behave unreliably. This timing check is often used in cells with multiple clocks.

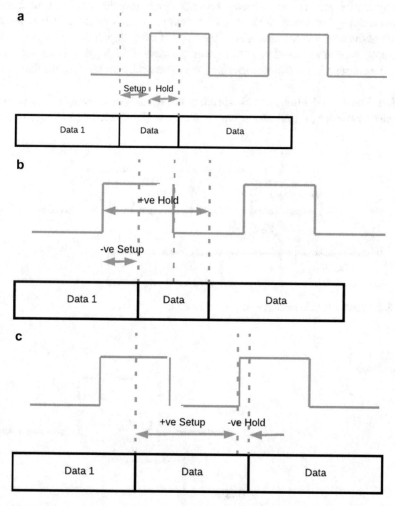

Fig. 5.20 Timing checks. (**a**) Positive setup positive hold. (**b**) Negative setup positive hold. (**c**) Positive setup negative hold

Setup and hold time checks are done with respect to the main control signal as in Fig. 5.21 where the data or address bus has to be stable. This check is done for embedded memories.

Removal Time: The removal timing check specifies a limit for the time allowed between an active clock edge and the release of an asynchronous control signal from the active state, for example, the time between the active edge of the clock and the release of the asynchronous reset signal for a flip-flops as shown in Fig. 5.22. If the release of the reset occurs too soon after the active clock edge, the state of the flip-flop becomes uncertain. The output q in the flip-flop can have the value set by the clear, or the value clocked into the flip-flop from the data input.

Recovery Time: The recovery timing check specifies a limit for the time allowed between the release of an asynchronous control signal from the active state of the next active clock edge as shown in Fig. 5.23, for example, a limit for the time between the release of the reset and the next edge of the clock of a flip-flop. If the active clock edge occurs too soon after the release of the reset, the state of the flip-flop becomes uncertain. The output q in the flip-flop can have the value set by the reset, or the data input.

Period The period timing check specifies the minimum allowable time for one complete cycle (or period) of a signal as shown in Fig. 5.24. The minimum period

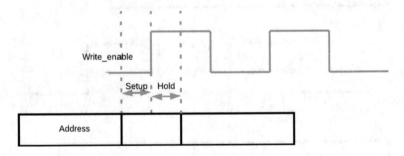

Fig. 5.21 Setup and hold timings of sequential elements

Fig. 5.22 Reset removal time

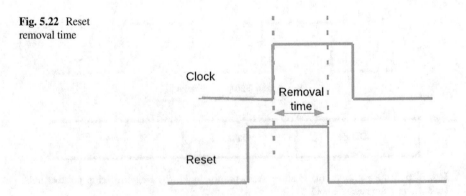

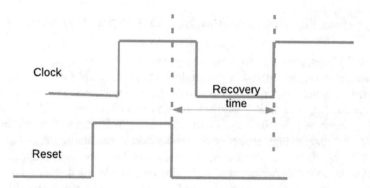

Fig. 5.23 Recovery time

Fig. 5.24 Clock period

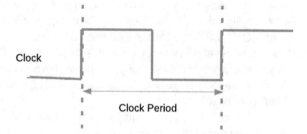

Fig. 5.25 MPH and MPL

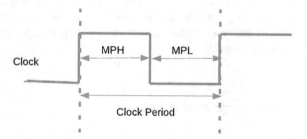

of the clock should be equal to maximum flip-flop propagation delay and maximum combination logic delay in a path for the design to work.

Minimum Pulsewidth Low (MPL) The MPL timing check specifies the minimum time a negative pulse must remain low. This timing check applies to "negedge" logic as shown in Fig. 5.25 Also it will be used for transparent latch setup and hold requirement used for slack adjustments.

Minimum Pulsewidth High The MPH timing check specifies the minimum time a positive pulse must remain high. This timing check corresponds to the positive edge logic.

5.10 Modeling Process, Voltage, and Temperature Variations

Process (P) conditions vary from one integrated circuit (IC) to another or die to die on a wafer. During the operation of a particular IC, the voltage (V) and temperature (T) can vary depending on the functional modes or slowly over time. At any instant in time, however, these variations are assumed to be small across a single IC. Usually, a timing library is characterized by a certain set of conditions: a particular process, voltage, and temperature. Based on the timing data in the timing library, the delay calculator reports pin-to-pin delays, interconnect delays, and timing check values. However, when the circuit operates under different conditions than those for which the library was characterized, the reported delay calculation values can differ from the actual values. To reflect the change in conditions, the delay calculator can scale the values. TLF uses models to define scaling factors (or multipliers) for PVT variations. Each multiplier is determined using the model and the actual condition value. For example, the multiplier to account for voltage changes is calculated from the model VOLT_MULT, which is a function of the voltage. Similarly, the process and temperature multipliers are calculated from the models PROC_MULT and TEMP_MULT, which are functions of a process variable and the temperature, respectively. The three multipliers are then simultaneously used to derate the delays and timing checks (Fig. 5.26).

The P, V, and T variables can be used for best, typical, and worst-case analysis and they can be specified in the form of triplets to reflect these cases. When the P, V, and T variables are in the form of triplets, the final derated delays are also in the form of triplets. In recent times, the SoCs have been integrated with sensors for monitoring process, temperature, and voltage abnormalities through on-chip circuits and sensors and control logic, which interrupts the processor to take action immediately, thus avoiding fatal errors.

5.10.1 Equivalent Cells

In some designs, identical cells are connected in "parallel" to increase drive currents, as shown below. For cells to be considered in parallel, all the identical inputs and outputs must be tied together, as shown in Fig. 5.27. Such configurations with identical cells can be recognized by the delay calculator so that they can be treated in a special way when doing delay calculations.

If cells are identical in behaviour but not physically identical (e.g., two buffers with different cells with different delay data or different drive strengths), some

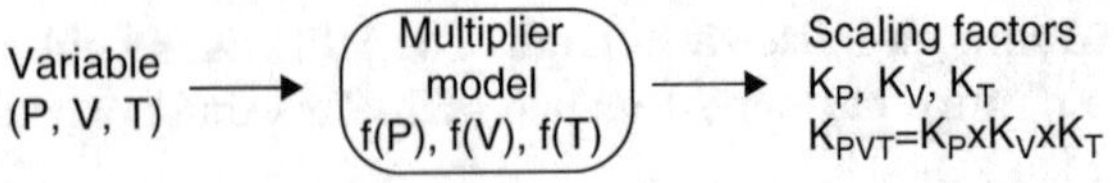

Fig. 5.26 PVT variations

Variable (P, V, T) $\longrightarrow$ Multiplier model $f(P), f(V), f(T)$ $\longrightarrow$ Scaling factors K_P, K_V, K_T $K_{PVT} = K_P \times K_V \times K_T$

Fig. 5.27 Equivalent cells

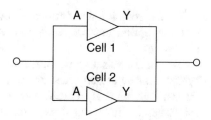

delay calculators require the cells to be labelled as equivalent in order to recognize them as being in parallel. Only with such labelling can those delay calculators recognize these cells as being parallel and make the improvement in drive strength. Additionally, the corresponding pin names of the cells must match. That is, for two dissimilar buffers, the pin names for both cells should be the same. In the example shown above, the input and output pins of both cell 1 and cell 2 are the same.

5.11 Timing and Design Constraints

Timing and design constraints describe the "design intent" and the surrounding constraints, including synthesis, clocking, timing, environmental, and operating conditions. Set these constraints on start points and end points to make sure that every path is properly constrained to obtain an optimal implementation of the RTL design. A path begin point is from either an input port or a register clock pin, while an end point is either an output port or a register data pin.

Use these constraints to:

- Describe different attributes of clock signals, such as the duty cycle, clock skew, and the clock latency.
- Specify input and output delay requirements of all ports relative to a clock transition.
- Apply environmental attributes, such as load and drive strength to the top-level ports.
- Set timing exceptions, such as multicycle paths and false paths.

In addition to specifying the timing and design constraints, one can specify optimization constraints. By default, the tools try its best to build logic to get the worst possible negative slack (WNS) numbers. To optimize, if the tool finds a WNS path that is meeting timing, then it optimizes the path with the next WNS. This continues until all paths meet their timing goals. However, the optimization process stops when it finds a path that is WNS and not meeting timing. Here the designer can specify the group timing paths into different cost groups. When multiple cost groups exist, the tool will optimize the WNS path in each cost group. If it cannot meet the timing goal for the WNS path in a cost group, then Genus will continue to try and optimize the WNS paths in each of the other cost groups.

A cost group is a set of critical paths to which you can apply weights or priorities that the optimizer will recognize. Paths assigned to a cost group are called path groups.

Timing analysis is carried out in two methods: one with wire-load models during synthesis or by actually feeding the layout information in the form of LEF files to the static timing analyzer (STA) to reduce the risk of timing closure after the physical design. Static timing analysis execution flow is shown in Fig. 5.28.

The purpose of timing analysis are to make sure the design meets the design goals after synthesis. Timing analysis identifies problem areas in the design and helps you

Fig. 5.28 STA command flow or tool flow

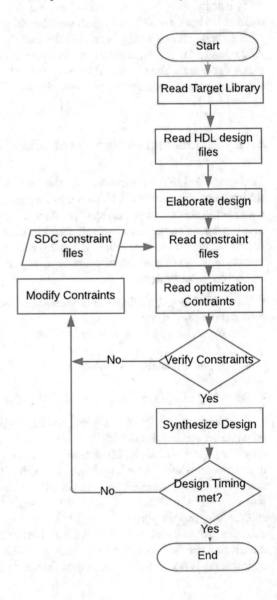

determine ways to solve these problems. After synthesizing a design, generate timing reports and area reports and check the selection of design elements in the design netlist. For static timing analysis, the design is represented as a set of timing paths. Paths are inputs to registers, register to registers, and regiister to outputs with start and end points. The actual data arrival time is computed and compared with the expected arrival time of data as specified in the constraint during analysis. That is, analyzing the timing compares the actual path delays with the required path delays of the design. Timing analysis computes gate and interconnect delay, traces critical paths, and then uses the critical path values to create timing reports. This helps you identify constraint violations. Constraint violations are negative slack in the path. It ensures that the setup and hold requirements of all the sequential elements in the design timing paths are met, or else by suitable algorithms, the violations are fixed by the concept of slack borrowing and slack stealing from cascaded paths by inserting transparent latches appropriately. The leftover paths which the STA tool cannot fix are to be handled by manually fixing the timing violations.

Figure 5.28 shows the timing analysis flow for the single functional mode of the SoC. If the SoC is designed for multiple modes, timing analysis is carried out for each of the modes and the timing violations must be fixed. The violations are fixed by replacing the standard cells from alternative choices in the library or by modifying the constraints in a few cases. In the worst case, the RTL design is modified to meet the required timing. Functional modes in SoC are controlled by a set of constraints during the mode in the design which drive timing analysis. A design has several functional modes, such as test, scan, and normal functional modes. For example, in a multiple supply voltage (MSV) design, a functional mode can have many states, such as shutdown mode, hibernate, and active modes. The timing constraints for each of the modes will be different and sometimes conflicting from one another. The design needs to be analyzed in each of the modes with appropriate constraints. Any violations found are appropriately fixed. The timing fix in one of the modes may introduce a critical path in another mode. Today's optimization tools support multimode timing analysis and optimizations, thus reducing by one extra design cycle.

5.12 Organizing Paths to Groups

Organize timing paths in your design into the following four cost groups:

- Input-to-output paths (I2O).
- Input-to-register paths (I2R).
- Register-to-register paths (R2R).
- Register-to-output paths (R2O).

Organizing paths in the design into groups is helpful when generating timing reports for analysis. The organization of the timing paths helps to break the number of path sets so that the designer can use his scripting skills to sort, find, and replace

Module: Adder
technology library: gpdk180nm
operating conditions: typical_case (balanced_tree)
wireload mode: zero

Pin		Type	Fanout	Load (fF)	Slew (ps)	Delay (ps)	Arrival (ps)
(Clock clk)		launch				0	R
a_ff0/clk	<<<					0	R
q_ff0/q	(u)	unmapped_d_flop	3	35.8	0	+120	120 R
add_1_2/A[0]							
g2/in_0						+0	
g2/z	(u)	unmapped_NAND2	1	13.6	0	+126	246 F
g4/in_0						+0	
g4/z	(u)	unmapped_NAND2	3	25.6	0	+156	402 R
g14/in_0						+0	
g14/z	(u)	unmapped_NAND2	1	25.6	0	+126	528 R
g26/in_0						+0	
g26/z	(u)	unmapped_xnor2	1	23.6	0	+136	664 R
add_1_2/z[4]							
z_ff3/d	<<<	unmapedd_flop_	3			+0	664 R
z_ff3/clk	(u)	unmapped_d_flop			0	+120	784 R
(Clock clk)		Capture					500 R

Timing Slack :   -284 ps (Timing Violation)
Start Point: a_ff0/clk
End Point: z_ff3/d
u: Unmapped pin(s).

Fig. 5.29 Timing report from synthesis tool

and strategize the way to fix the violations. By default, the timing report shows the critical path from each path group. The critical path is the timing path in the design with the greatest amount of negative slack (margin). The goal of the designer to have all design paths with positive slacks, with enough margin. This extra margin is to balance for any error between the STA design algorithms and actual design timings when fabricated. Fixing the timing violation involves standard cell replacements with better propagation delays and registering the intermediate cell in the path, thus breaking it into two paths without affecting the functionality and getting the waver if the path is a false path. A typical timing report is shown in Fig. 5.29.

As it can be seen, the path is register-to-register (R2R) path with a start point as *a_ff0/clk* and an end point as *z_ff3/d*. The instance u contains unmapped pins, with a negative slack of 284 ps. The path consists of *d flip-flop* and *nand2* and *xnor2* cells. The violation in the path can be fixed by two ways: by changing the *nand2* and *xnor2* cells to faster cells if they are available in the standard cell library and by splitting the path by registering the output of second *nand2* if it does not affect the functionality. If the path is split by registering the output of the second *nand2* cell, new path will terminate at another *d flip-flop* which will be the end point of the new R2R path, and the new path timing would be 402 ps. With the capture timing of

500 ps, it will result in positive slack. However, the effect of this change on functionality will be then verified by running logic equivalence with the modified netlist and the golden reference RTL file.

5.13 Design Corners

Design corners represent the behaviour of the design at different process, voltage, and temperature conditions. The process, voltage, and temperature (PVT) variations and their effects on the transistors are modelled as PVT models of the transistor as shown in Fig. 5.31. The technology library is referred to by the transistor channel lengths L. For example, 45 nm technology has a transistor channel length of 45 nm, and 65 nm technology has the transistor channel length of 65 nm. The process represents the length L of the transistor. For the same temperature and voltage, the current will be more in 45 nm technology than of 65 nm technology owing to the formula $I = \mu C_{ox} \dfrac{W}{L} \left(V_{gs} - V_t^2 \right)$. Recalling the transistor theory, the smaller the process L, the larger the current. This current will charge and discharge the capacitor faster, and hence the delay will be less (Fig. 5.30).

The supply voltage is fed to the SoC from outside or through the on-chip regulators. This voltage can change over time. Hence, SoC is designed to work accurately

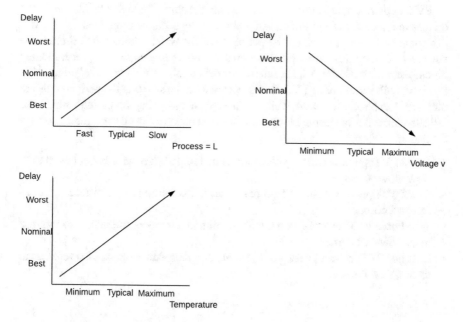

Fig. 5.30 PVT characteristics of transistors

Fig. 5.31 Multimode
timing constraint analysis
script file

```
set_attribute lib_search_path ...
set_attribute hdl_search_path ...
set_attribute library ...
read_hdl <hdl_file_name>
elaborate
create_mode -name {function Sleep Test}
read_sdc -mode function SOC_main.sdc io_function.sdc
read_sdc -mode Sleep SOC_sleep.sdc io_SOC_sleep.sdc
read_sdc -mode SOC_Test SOC_test.sdc io_test.sdc
syn_map
report timing
write_encounter design
```

over a range of voltages with the typical voltage of the claimed voltage in the data-sheet with ±10 variation. From the equation mentioned above, the higher the voltage, higher the current, and the faster the circuits.

The SoC circuit operations also depend on the ambient temperature. The higher the temperature, the higher the electron collision in the device and the current flow reduces and hence the delay increases. This also needs to be modeled and SoC is expected to work accurately in all working environments at all prevailing ambient temperatures.

PVT modelling captures this variation within the chip. The logic circuits fabricated on dies in the center of the silicon wafer show pretty accurate properties in PVT values compared to the circuits on the periphery of the wafer. Though the difference is not much, it can affect the logic functionally. This is modelled as a process called on-chip variation (OCV). So, the inter-chip variations of PVT are modelled as OCV and intra-chip variations as PVT. It is expected to make sure that the design goals are met with these variations also. This is achieved by analyzing the timing using these models. Some normal terminology used in the context of SoC design timing are the following:

- Worst PVT: process worst, voltage min, and temperature max, also referred to as slow-slow corner.
- Best PVT: process best, voltage max, and temperature min, also referred to as fast-fast corner.
- Worst cold PVT: process worst, voltage min, and temperature min, also referred to as slow-fast corner.
- Best hot PVT: process best, voltage max, and temperature max, also referred to as fast-slow corner.

5.14 Challenges of STA During SoC Design

SoCs of today operates in multiple modes like active, sleep, and test modes, to name a few, and the timing requirements in each of these modes are different. Mode is a set of functional behaviour of the system. These modes share the same logic in many places in the design. It is required to meet the static timing in all these modes separately for reliable operation of the system. Static timing analysis will require a different set of design constraints in each of these modes. For example, the design in sleep mode may use a different supply voltage or system clock frequency. Fixing the timing issues in one mode may result in issues opening in the other mode, thus contradicting the design needs. To take care of these contradictions, the static analysis tools support multimode timing analysis capability. This involves creating modes in the constraint files and feeding corresponding constraint files for generating reports. The violations in the reports are fixed by the same method as the issues in the single mode SoCs. A typical STA analysis script file for multimode SoC is shown in Fig. 5.31. In the example shown, the SoC is functioning in two modes apart from the normal active mode. They are sleep mode and test mode, and the corresponding constraint files are read into the STA analysis tool in the script. The tool picks appropriate timing data from the technology library for analysis and generates timing reports.

SoC design timing is affected by multiple parameters like the routing delays, load on the logic, and the fanouts of the gates used. Any design change which is inevitable during the design phase may result in different paths, which can be seen in multiple runs of the STA reports. Hence, it is a continuous process to perform STA timing analysis till the design is finalized. Apart from the timing reports for analysis, the reports also point out the un-clocked registers, multiple driven registers, combinational loops, and redundant logic that has to be corrected knowing the design details. The STA tools also have the capability of identifying these in the SoC design to help the designers fix them.

Chapter 6
SoC Design for Testability (DFT)

6.1 Need for Testability

The functionality of SoC is guaranteed if its design and fabrication are done correctly. Verification during the design process confirms the correctness of design. But testing the fabrication of millions of transistors, which make the device, is impossible to verify on the chip. Complicating the problem further is the complexity of SoC design, which is increasing day by day. The testability after fabrication is an important factor for its success. Design for testability (DFT) is an important practice, that provides a means to comprehensively test a manufactured SoC for quality and coverage. Failures to detect flaws in fabrication before putting a chip in a product can be disastrous and often fatal. In SoCs, all transistors and internal interconnects of the SoC design are generally inaccessible to test even during fabrication to ensure correct fabrication. Special techniques are required to make device testing possible. DFT is the concept of adding extra logic during the design to enable testing of most of the logic design. To ensure correct fabrication, all sequential cells such as D flip-flops, memories, and input-output pads, which are generally inaccessible, to ensure correct fabrication need special logic to test either directly or indirectly. In the majority of the SoCs, approximately 70 to 75% of the logic circuit is comprised of sequential design elements such as D flip-flops. Around 60% of the majority of the SoC's silicon area is on-chip memory. Input-output pads in SoC design provide access to the outside world in products. Hence, if all these design elements are made testable by some means, there is a high probability of SoC working. However, to achieve this, it is needed to get one hundred percent coverage on testability of design elements. The DFT techniques aim to achieve this goal. These techniques need additional test mode to be added to the SoC design with extra logic which increases the SoC design area. Considering the criticality of the problem and the exorbitant cost of respin of SoC designs, it is still a requirement. Figure 6.1 shows the DFT steps in SoC design flow.

© The Author(s), under exclusive license to Springer Nature Switzerland AG 2022
V. S. Chakravarthi, *A Practical Approach to VLSI System on Chip (SoC) Design*,
https://doi.org/10.1007/978-3-031-18363-8_6

6.2   Guidelines for SoC Design for Testability

Most of the SoC designs are synchronous sequential logic functions with predictable behavior. Therefore, they are inherently testable. It is easy to implement test logic around synchronous logic to ensure correct manufacturability. But with the functional complexity, multiple complex clocking schemes pose challenges in making a chip testable. It is necessary to follow the DFT guidelines to make a chip testable for manufacturing issues. The following are the design guidelines for testability of SoC designs:

- The system architecture must have minimum number of clocks most preferably single clock or clocks required for other subblocks in the design generated from a common clock source.
- All the inputs must be registered (stored in registers before processed) to avoid signals leading to metastability at the point of processing them.

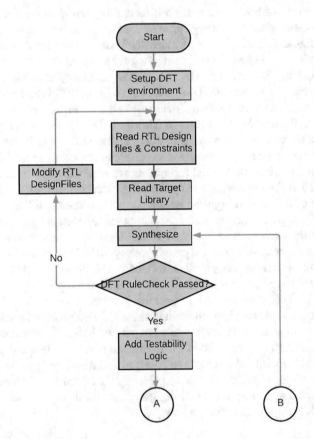

Fig. 6.1 DFT flow in SoC design

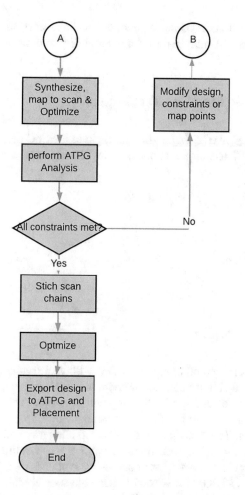

Fig. 6.1 (continued)

- Control signals such as set, reset, and clock inputs of the flip-flops should not have any combinational logic in their paths as much as possible.
- Asynchronous signals for reset input of the flip-flop must be avoided if possible.
- None of the clock inputs of the timing elements in the design are to be gated or delayed through delay cells or buffers.
- Input signals must not be delayed through delay cells.
- Consider routing net delays are to be shorter than logic propagation delays of the cells.

Though most of the logic blocks are synchronous, there will be a few asynchronous blocks that pose huge challenges for testability. It is good practise to separate the asynchronous logic block and isolate it from the synchronous SoC logic core for easy implementation of DFT logic. Spreading the asynchronous logic all around the

SoC design makes it not testable. Whenever the above rules are violated in a SoC design, it is essential to analyze for timing, testability, and manufacturability.

6.3   DFT Logic Insertion Techniques

DFT techniques involve adding additional logic to the SoC design to make it testable. The main DFT techniques adopted during the SoC design are:

- Scan insertion.
- Boundary scan.
- Memory BIST.
- PTAM.
- Logic BIST.
- Scan compression.
- OSCG.

6.3.1   Scan Insertion

Scan insertion is the method of replacing D flip-flops of the SoC design with scannable flip-flops and serially connecting them into a chain as shown in Fig. 6.2. A scannable flip-flop cell is a special flip-flop with test logic. This makes most of the SoC design testable, as in most of the synchronous SoCs, around 70% of the design cells are flip-flops. The extra scan logic inserted allows you to test the sequential state of the design through the additional test pins of the scan flip-flops in test mode. Scan cell insertion and stitching are done using synthesis tool. The automatic test pattern generator (ATPG) tool is used to generate special scan patterns to test the design by means of fault simulation. The test patterns are test vectors which are fed into the design through test input pins and by capturing the design response at the scan test outputs. The fault in the chain is identified and fixed by a scan test during design. The goal of the ATPG is to achieve higher fault coverage and generate a more compact test pattern-set for the design. The scan test concept is shown in Fig. 6.2. As in the figure, the scan_input (scan_in) and scan enable (scan_en) are extra scan test input signals, through which scan chains are loaded with test patterns. Scan chains are not required in the normal functional mode. To enable the scan mode for DFT, the scan mode (scan_mode) input signal is added and the output response of the scan chains in the design is monitored at the output signal scan output (scan_out). Hence, scan chains in SoC design have separate input and output access for DFT. During scan-mode, test data is shifted through the scan chains. There can be as many sets of scan test input-output pins as chains. The scan test pattern is shifted through the scan_in input pins and shifted out through the output scan_out pins. These extra input-output pads can be multiplexed with the

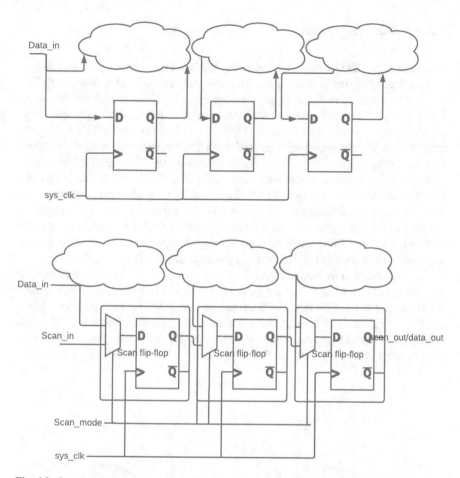

Fig. 6.2 Scan insertion concept

functionally compatible input-output pads without increasing the number of IO pads in the SoC. The length of the scan chain depends on the memory capacity of the tester, which can hold the test pattern. In practice, there will be around 2000–2500 scan flops connected in a scan chain. Hence, depending on the complexity of the SoC, the number of scan chains is decided, and accordingly, scan in and scan out signals will scale up. The control signals like scan mode and scan_en are shared across chains in a SoC.

To insert the scan chain, it is required to check if the D flip-flops are all testable and the clocks are controllable. It is also required that the asynchronous/synchronous resets be held at inactive levels in scan test mode. These are checked as a process called DFT rule check during the SoC design. There are some lint tools which check the design for DFT rules.

6.3.2 Boundary Scan

Boundary scan (BS) logic is inserted to test the input-output pads of a SoC design, independent of its functionality. Boundary scan cells are inserted between each SoC pad and the system functional logic. They are then connected at the boundary, similar to scan chain, called a boundary register chain. The entire boundary scan logic inserted has to comply with the IEEE 1149.1 or 1149.6 standard which defines the procedure to test the input-output pads of the SoCs. The boundary scan test insertion consists of the insertion of the JTAG macro core, insertion of the boundary scan cell, and connecting them as a boundary scan chain. The JTAG macro core is inserted into the netlist manually or as a part of the boundary scan insertion procedure. The JTAG macro is a generic core used for interconnect testing on printed circuit boards by monitoring the value of each chip's input and output independent of on-chip system logic. The JTAG core enables controlling the pattern in and out of the boundary scan register for testing. The boundary scan concept is shown in Fig. 6.3.

As it can be seen in Fig. 6.3, the BS cells are added in between the SoC IO pad and the system core logic. The BS cells are connected to form a chain of registers which are fed by the JTAG core with the test pattern. When the pattern is completely shifted, it is shifted out through the test output pad, which is monitored. This test pattern can also be bypassed and sent directly to the test output pad to test the IO pads of other chips on the board. The JTAG core has five standard IO ports called:

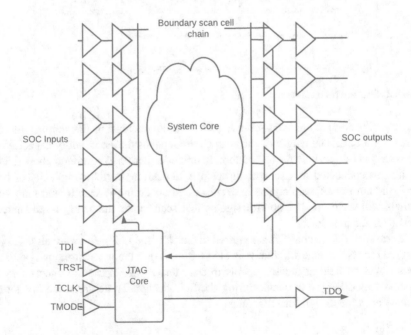

Fig. 6.3 Boundary scan concept

- Test data input (TDI): Input port through which the test pattern is fed in.
- Test clock (TCK): Test clock used to test the IO pads.
- Test mode select (TMS): When set, enables the pad testing through boundary scan logic.
- Test reset (TRST): Optional test reset input port to reset the test logic and state machine.
- Test data output (TDO): Output port through which the pattern can be monitored.

The standard JTAG core has to be compliant with the IEEE Std. 1149.1 standard and the later IEEE Std. 1149.6 for test access port (TAP) and boundary scan architecture targeting manufacturing faults in the SoC ports and in the interconnection on PCBs. The JTAG core logic in the boundary scan architecture is shown in Fig. 6.4.

A standard JTAG core logic inserted as the boundary scan test logic contains:

- Test access port (TAP) controller which is the control state machine generating control signals to various internal logic.
- Instruction register (IR) which holds the opcode of the test instruction to be processed.
- Instruction decode logic which decodes the instruction written into the instruction register.

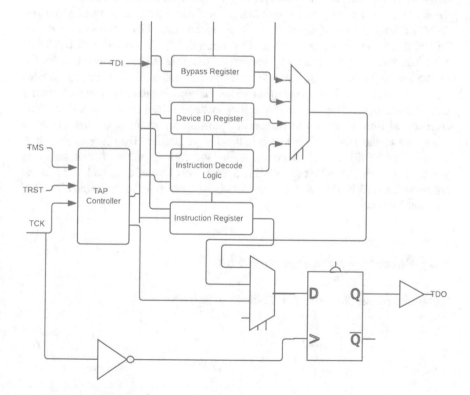

Fig. 6.4 JTAG BS architecture

- Bypass register (BR) which blocks the test pattern to be fed to the boundary scan chain but passes the pattern to the TDO port.
- Device ID register which holds the unique identification number of the SoC device.
- Test data output (TDO) which outputs the test pattern after it is adequately shifted through the BS chain.
- (Optional) Custom test data registers to support user-defined test register which enables custom test to be done on the IO pad specific to the SoC. This is not necessary but optional facility provided to the designer.

To test the IO pads, instruction code is fed through the TDI pin into the instruction register of the JTAG core. The instruction is decoded, based on which the data register with the identified pattern is shifted through the chain of Boundary scan (BS) cells by feeding as many clock pulses equal to the number of BS cells and is shifted out through the TDO output pin to get the same pattern as input. This ensures that the pads are working as intended. It is required to support the four instructions: BYPASS, with instructions EXTEST, RELOAD, and SAMPLE when the JTAG core is used. The mandatory instructions ensure the SoC chip interface test on the PCB is doable. The BYPASS test is done to bypass the internal boundary scan register and access the next chip interfaced to the SoC chip under consideration, while the EXTEST is the external test by feeding the desired pattern. The RELOAD and SAMPLE tests are user-defined. In addition to these tests, JTAG supports accessing DEVICE ID and DATA registers in TAP through ID_CODE and USER_CODE tests. The designer can insert any number of the data registers, supported by the multiplexer logic, to choose one among them. The TAP controller FSM generates a control signal for selecting the data register and shifting the data pattern from the data register depending on the instruction loaded in the instruction register. The selection of instruction or test pattern and shifting the result of the instruction is done through the TDI and TDO ports in the design. Earlier JTAG core compliant to the standard IEEE 1149.1 does not define the testing of the differential pads and the interconnects with capacitive coupling. This limitation is addressed in the later standard IEEE 1149.6. For more details, you can refer to the respective standard documents.

6.4   Boundary Scan Insertion Flow

The boundary scan insertion flow is shown in Fig. 6.5.

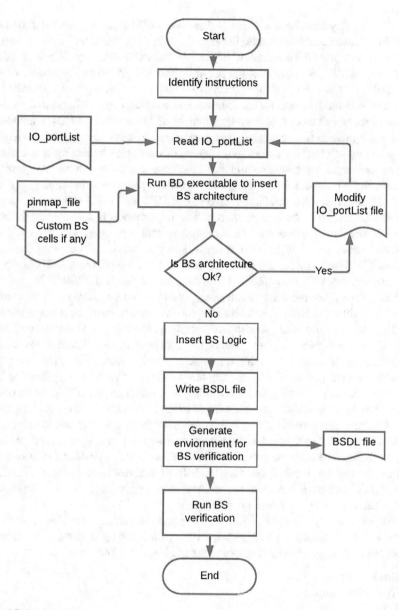

Fig. 6.5 BS insertion flow

6.4.1 Memory Built-in Self-Test (MBIST)

Embedded memories in a SoC are tested by self-test structures called memory built-in self-test (MBIST). One or more MBIST structures are added to memory behaviour models. Hence, this can be directly instantiated into the SoC design. The

MBIST circuitry interfaces with the higher-level SoC functional blocks of the system. In system functional mode, through the interface, functional system data is passed to the embedded memory, bypassing the BIST circuitry. When in BIST mode, the MBIST circuitry runs the self-test function, providing a signature-based pass/fail and "test complete" indication to the system, which can be accessed by the user. The self-test function for the memory can be modelled as a behaviour models using HDL, which can be verified by simulations using standard HDL simulators. The BIST architecture can also be customized in many cases, which enables the grouping of small memories into clusters of memories, executing user-defined test patterns, and generating customizable address sequences, for memory testing. Today's SoCs contain a large number of embedded memories, and testing of them needs an automated test strategy. Conventional DFT and ATPG approaches cannot be used for testing embedded memories. The fault models of memory differ from those of standard logic design fault models in that memories will have address faults, memory cell faults, retention faults, stuck-at faults, and coupling faults, to name a few. Furthermore, using external automatic test equipment (ATE) to apply test patterns targeting these faults is also impractical and inefficient as large numbers of patterns are required to test every memory cell structure and also cannot cover all faults. Controlling and observing each memory from the primary pins of the SoC requires too much silicon real estate and reduces the performance of the SoC. If test patterns are applied from an external source, they cannot be reused for the next generation of SoCs using the same memories. These limitations are overcome by integrating an MBIST architecture involving a test pattern generator and response comparator logic into the SoC design. Advantages of MBIST are that SoC testing can be done without the need for an external tester and can be done as functional testing, thus providing a test mode. With on-chip pattern generation circuitry, the test is executed so fast and with a signature-based response analysis and generating result that it reduces the need for an external analyzer and external data storage. Hence, the test overhead of inserting the MBIST architecture into the SoC is very less. BIST integration is similar to any other functional block integration. The MBIST architecture is shown in Fig. 6.6.

SRAM memory consists of three main parts: an address decoder, a memory array, and the memory access logic. A memory fault can be in any one of these or more MBIST targets. Major memory faults are classified into:

- Stuck-at faults.
- Transition faults.
- Coupling faults.
- Pattern-sensitive faults.

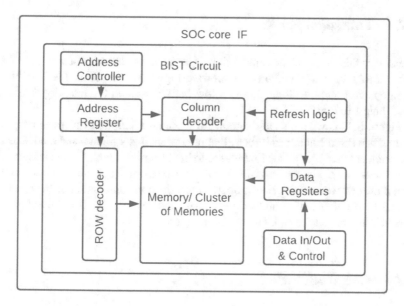

Fig. 6.6 MBIST architecture

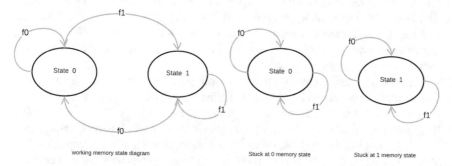

Fig. 6.7 Stuck-at fault state diagram

6.4.2   Stuck-at Faults

Memory control logic or array appears to be stuck at one logic level either 1 or 0. This is called *stuck-at fault*. Stuck-at faults model this behavior, as a signal or cell appearing to be tied to power (stuck-at-1) or ground (stuck-at-0). Figure 6.7 shows the state diagram for a stuck-at fault.

To detect stuck-at faults, it is necessary to force the value opposite to that of the stuck-at fault at the fault location. For example, to detect all stuck-at-1 faults, it is required to drive 0 s at all fault locations. To detect all stuck-at-0 faults, it is required to force 1 s at all fault locations. BIST patterns generated internally for self-test will generate such patterns and drive the memory circuit.

6.4.3   Transition Faults

A memory fails if any of its control signals or memory cells cannot transition from either 0 to 1 or 1 to 0. Figure 6.8 shows a high-transition fault, the inability to change from logic 0 to logic 1, and a low transition fault, the inability to change from logic 1 to logic 0.

Figure 6.9 shows the state diagram for a memory cell that functions correctly when it is written 1 and read back 1. Test pass when it is written 0 and read 0, as the transition is from 1 to 0. Due to its "zero to high transition fault," when it is written with 1 and read again, the test fails. However, a stuck-at-0 test might not detect this fault if the cell was at 1 originally. So, to detect the transition fault, it is to be written 1, read 1, written 0, read 0, and written 1 again and read. If it reads 1, the test passes, or else it shows that the cell has transition failure.

6.4.4   Coupling Faults

Memories also fail when a write operation in one cell influences the value in another cell. Coupling faults model this behavior. Coupling faults fall into several categories: inversion, idempotent, bridging, and state. Figure 6.10 shows that inversion coupling faults, commonly referred to as CFins, occur when one cell's transition causes an inversion of another cell's value. For example, a 0-to-1 transition in cell_n causes the value in cell_m to invert its state.

Figure 6.11 shows that idempotent coupling faults, commonly referred to as CFids, occur when one cell's transition forces a particular value onto another cell. For example, a 0-to-1 transition in cell_n causes the value of cell_m to change to 1 if the previous value was 0. However, if the previous value was 1, the cell remains at 1.

Bridge coupling faults (BFs) occur when a short, or bridge (low strength connection due to metal deposit or polysilicon connection), exists between two or more cells or signals. In such a case, a particular logic value triggers the faulty behaviour

Fig. 6.8   Transition fault

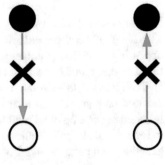

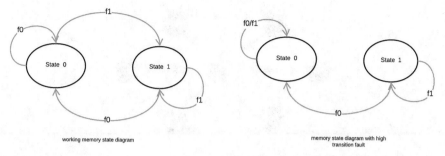

Fig. 6.9 Stuck-at fault 0 memory state machine

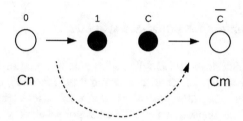

Fig. 6.10 Inversion coupling fault

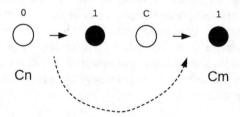

Fig. 6.11 Coupling fault

rather than a transition. Bridging faults fall into either the AND bridge fault (ABF) or OR bridge fault (OBF) subcategories. ABFs exhibit AND gate behavior; that is, the bridge has a 1 only when all the connected cells or signals have a 1. OBFs exhibit OR gate behavior; that is, the bridge has a 1 value when any of the connected cells or signals have a 1 value. State coupling faults, abbreviated as SCFs, occur when a certain state in one cell causes another specific state in another cell. For example, a 0 value in cell i causes a 1 value in cell j. Coupling faults involve cells affecting adjacent cells. Therefore, to sensitize and detect coupling faults, "March tests" perform a write operation on one cell (j) and later read the cell (i). The write/ read operation performed in ascending order of address detects a coupling fault of the addresses. This marching is repeated even in ascending addresses.

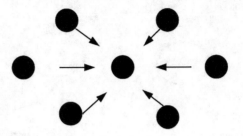

Fig. 6.12 Neighborhood pattern-sensitive fault

6.4.5   Neighborhood Pattern-Sensitive Faults

Another way in which memory can fail is when a write operation on a group of surrounding cells affects the values of one or more neighbouring cells, as shown in Fig. 6.12. Neighbourhood pattern-sensitive faults model this behavior. Neighbourhood pattern-sensitive faults break down into three categories: active, passive, and static.

An active fault occurs when, given a certain pattern of neighboring cells, value change in one memory cell causes a change in the value of the other memory cell. The effect of change on the neighbouring memory cell due to writing a value in a particular memory cell can create different kinds of faults. If the effect is to fix the value of a memory cell to a particular value, then it is called a passive fault or static fault. This effect can be so complex that the detection of these faults becomes equally difficult and requires multiple special sets of algorithms to generate test patterns to detect them. This opens the way for ongoing research to arrive at a variety of algorithms to detect these faults.

6.4.6   MBIST Algorithms

There are memory test algorithms that can drive patterns to detect the commonly occurring faults in memories. Many of these algorithms are implemented as logic which generates the patterns and can test multiple on-chip memories. The most commonly used algorithms are the March algorithms. There are many algorithms used in MBIST, like advanced test sequence (ATS), walking 1 or 0 s, March A and March B, March C, and checkerboard.

The March C algorithm detects the following multiple faults:

- Stuck-at.
- Transition.
- Coupling—unlinked idempotent and inversion, and other coupling faults on bit-oriented addresses.

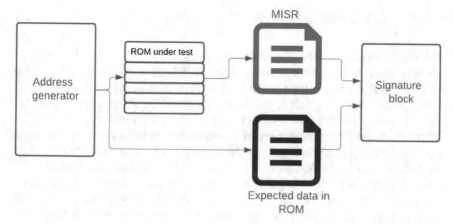

Fig. 6.13 ROM test algorithm

6.5   ROM Test Algorithm

The ROM test algorithm provides address and control circuitry fault detection. This algorithm reads the values from each address of the memory in increasing order, one word at a time, as shown in Fig. 6.13. To determine the pass/fail state of the memory, the circuit inputs the values read from memory into a multiple input signature register (MISR) and compares the signature against the known good value for the ROM.

Programmable memory BIST (MBIST) insertion is the process in which memory BIST logic is inserted that allows for control, testing, and diagnostics of the memory cell instances via IEEE 1149.1 or 1149.6 JTAG control or direct pin access control. Programmable memory BIST logic permits memory cells in the SoC independently from system modes. Insertion of the PMBIST logic is customised for each design using a configuration file.

6.6   Power Aware Test Module (PATM) Insertion

PATM insertion inserts overriding control logic into the design's power-manager control block(s) in order to stabilize the power-manager control pins to the switchable power domains during test. PATM logic is inserted into the design's power-manager control block(s) for the power domains defined in the UPF file. These are used to generate patterns for self-testing. This reduces the dependence on external automated test equipment (ATE).

6.6.1   Logic BIST Insertion

Logic BIST similar to MBIST, permits self-testing of SoC logic structures without the need for ATE. It involves insertion of the BIST logic, which automatically generates a pseudorandom pattern generator (PRPG), which is also called a shift register sequence generator (SRSG), and the multi-input signature generator (MISG) to capture the response of the patterns fed into the core. It is essential to ensure that the PRPG and the MISR generators generate unique patterns by suitably using the right set of generator polynomials and initialization sequences. The basic architecture of the LBIST is shown in Fig. 6.14, which is also called "self-test using MISR" and parallel SRPG (STUMP). The pseudorandom pattern generator (PRPG) generates a pattern which is shifted into the scan chains, and the patterns that are output through the scan chains are compared with the generated pattern, and pass-fail status is indicated through signatures. The signature can be read out by the direct access interface or through the JTAG TDO lines. Depending on the requirements for a SoC, both or either options can be provided to perform the LBIST test on a SoC.

The JTAG-based LBIST uses support for two instructions: RUNBIST and SETBIST as defined by IEEE 1149.1. The RUNBIST command uses internally generated patterns which are fed into the scan chains and the results are shifted out of the scan chains, to the MISR generator, which generates the signatures for multi-input sequences it gets from the scan chains. This MISR signature is either read out of the TDO line of JTAG or through direct access to the external pattern reader circuit. The difference between RUNBIST in JTAG mode and direct access mode is the external interface. The RUNBIST instruction, an 1149.1 IEEE instruction, enables the LBIST process. When RUNBIST is loaded in the instruction register (IR), the TAP controller state machine initiates the BIST process. RUNBIST acts as a select line. RUNBIST enables data to enter the SoC core from the BIST controller's PRPG, thus allowing the shift counter's value to control the shifting of the data through the STUMPS channels as shown in Fig. 6.15.

The shift counter begins at a state of all zeros. When RUNBIST executes, it counts upward until it reaches a specified limit corresponding to the length of the longest STUMPS channel. Each time it increments, data in the STUMPS channels shifts. Upon reaching this limit, the STUMPS channel data shifting stops and the BIST circuitry disables the scan enable line. This allows capture of system data in the scan cells. The shift counter then resets again to all zeros. It repeats this process for each pattern the PRPG applies to. Each time the shift counter resets to 0, it signals the pattern counter to decrement its value. When the RUNBIST instruction executes, the BIST controller loads the pattern counter with the number of patterns that the PRPG is to generate. Each time the shift counter resets to 0, the pattern counter is decremented by one. When the pattern counter reaches zero, this indicates that the PRPG has finished generating and applying patterns. To follow RUNBIST instruction rules, a zero value in the pattern counter triggers the BIST controller to disable the LFSR clocks. This ensures a stable final MISR signature in a situation where tests running simultaneously on different chips require different numbers of patterns for testing.

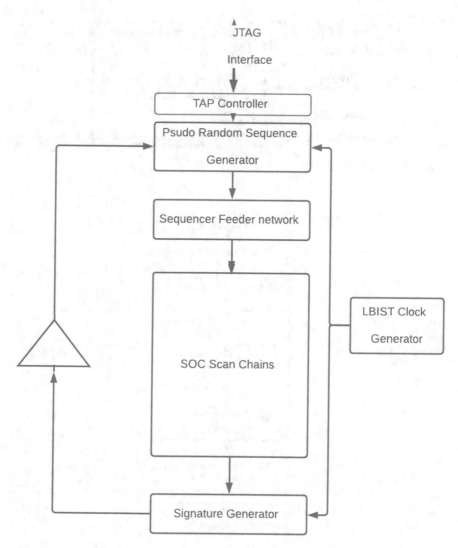

Fig. 6.14 LBIST architecture

The direct access interface will contain reset and an enable/disable port for LBIST. It uses the same JTAG macro for the tap controller functionality as the instructions defined in the JTAG macro. The SETBIST instruction permits the feeding of an externally generated pattern of choice based on the requirement. LBIST test function requires a LBIST clock generator for shifting out the patterns. One has to keep in mind the need to include compression logic to minimize the area overhead due to the LBIST logic. The standard DFT tools support adding the LBIST circuitry to the SoC. The LBIST insertion flow is shown in Fig. 6.16.

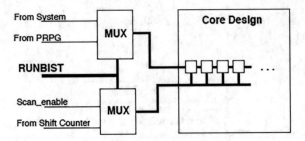

Fig. 6.15 RUNBIST function

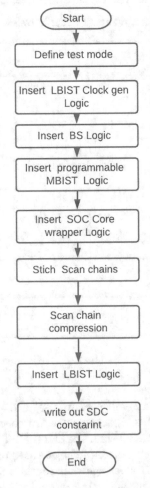

Fig. 6.16 LBIST insertion flow

6.6.2   Writing out DFT SDC

DFT SDC involves the writing of three types of constraints from the DFT phase of the SoC design. They are: SDC file with DFT mode disabled (NON DFT MODE), SDC constraint with DFT mode shift where the test patterns are shifted (DFT SHIFT MODE), and SDC constraint for capturing the response patterns from DFT logic (DFT CAPTURE MODE). It is essential to verify all the three constraints before they are finally used for DFT verification or synthesis.

6.6.3   Compression Insertion

The length of the scan chain poses a limitation on the depth of the test pattern to be held in ATE. In practice, the scan chains will be around 2000 flip-flops per chain. Today's SoC will have multiple scan chains to cover all the sequential elements. The test time on ATE is proportional to the number of scan chains and the number of scan cells in each chain. Hence, it is always preferred to adopt techniques to reduce the test times. The famous techniques adopted to reduce the test time are the insertion of compression logic to build internal scan channels, thereby reducing the ATE test times and the test pattern sets used to verify the design. Scan compression builds shorter internal scan channels from the top-level scan chains, thereby reducing the ATE test times and test data volume of the pattern sets. The compression logic is inserted as a compression macro with additional scan-multiplexing logic to define the internal scan channels.

6.7   On-SoC Clock Generation (OSCG) Insertion

Scan test is generally conducted at a very low frequency compared to the operating frequency of the SoC, which will be very high in the order of hundreds of MHz to multiples of GHz generated by a PLL internally. Though low-frequency tests get passed, there is a possibility of the logic failing at the operating frequency of the SoC. Feeding high frequency from external signal generating sources to the SoC for testing at the actual operating frequency is not possible because of the limitation of the normal pads, which cannot pass high-frequency signals. A concept called "at-speed" testing is adopted to test the SoCs at their operating speed. This involves insertion of the on-SoC clock generation (OSCG) logic. This avoids the additional expense and trouble of supplying high-speed clock signals from the automatic test equipment (ATE) and the use of special differential pads for the SoC. Typically, today's SoC contains PLL modules that generate high-speed clocks internally. The inserted OSCG logic is programmable to allow a certain number of these high-speed pulses from the on-chip PLL to be applied to the clock domains being tested using delay test patterns.

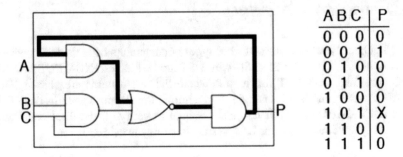

A B C	P
0 0 0	0
0 0 1	1
0 1 0	0
0 1 1	0
1 0 0	0
1 0 1	X
1 1 0	0
1 1 1	0

Fig. 6.17 Combinational loop

6.8   Challenges in SoC DFT

Today's SoCs impose many challenges for testability due to their special features and design styles. As asynchronous design blocks are not fully testable, most of the design styles using basic synthesis algorithms with standard cells and the FPGA architectures require synchronous design style to ensure that they are testable. Synchronous designs are more predictable. During standard gate array designs, synchronous design is enforced as coding guidelines to ensure that they are testable. To ensure design for testability, there are commercial tools available which, through a set of design rules, check the design and pop out violations. These tools ensure that the design is testable, manufacturable, and predictable in terms of functionality. It is based on the scan ability test run on synchronous designs. Design containing loop logic generally poses testability challenges. If the output of a combinational logic circuit is fed back to one of the inputs, it is termed a combinational loop, as shown in Fig. 6.17. Such circuits in the design netlist are a result of bad RTL coding practices. Such paths are to be broken by registering the output so that the combinational loop is broken and yet the functionality is met.

If the feedback path, that connects output signal to the input signal of any part of the circuit, passes through a sequential element like a flip-flop or latch, it is called a sequential loop. The design blocks with sequential loops are not testable for manufacturing defects using DFT techniques. The tools which test the testability of the design identifies such structures from the RTL models and issue errors and warnings to indicate that they are not testable.

6.9   Memory Clustering

SoCs typically has many memories of different sizes distributed in different modules. It is possible to add common MBIST structure to a group of memories by clustering them if they are of the same type, operate on the same frequencies, and are physically located close to each other. This helps to save the DFT overhead in terms of silicon area.

6.10   DFT Simulations

Once the DFT logic is inserted, it is necessary to verify the inserted logic and test mode functionality like the boundary scan, the scan tests through JTAG, and the BIST on memory and logic. Most of the commercial DFT tools write out the test environment, the test patterns, and the run scripts for running simulations and verification. They use fault models for generating test patterns to propagate and derive the expected responses from the DUT. If the expected response is not detected, the failure is flagged and the time is recorded. The same patterns are used for deriving ATPG vectors.

6.11   ATPG Pattern Generation

Once the DFT rule checking passes, the design with scan chains is fed to the ATPG tool to generate the test patterns. Design rules for DFT typically confirm that the scan patterns fed into scan chains are shifted out of scan outputs properly. If there are multiple scan chains, they are shifted out in parallel simultaneously. This is called the parallel scan test. The test patterns generated for running DFT simulations for scan and boundary scan are to be converted to a special format to enable regeneration as test patterns from automatic test equipment (ATE) in waveform generation logic (WGL), which is an ASCII file used to extract the waveform and edit and plot the information from waveform database (WDB). The test patterns in WGL format are required to test the fabricated dies using the testers at wafer and chip level.

6.12   Automatic Test Equipment Testing (ATE Testing)

Conventional DFT testing uses external test patterns in WGL format as stimulus, and an automatic way of applying a set of the patterns in a controlled manner to the SoC, knowing its IO pins and their physical locations and other details by the tester. The tester examines the device's response, comparing it against the known good response stored as part of the test pattern data. The effort at this stage is always to reduce the ATE test times by optimizing the test patterns but still sorting only the good chips from the lot. Some of the popular methods used to minimize the test times are at-speed testing using an on-chip high-speed clock for testing, scan compression, and self-testing methods.

Chapter 7
SoC Design Verification

7.1 Importance of Verification

First-time success and field success are absolute requirements of SoC designs, because of the exorbitant cost of development and fabrication. It is verification at all stages of design and development, that guarantees this. For decades, people have worked to make SoC verification effective and efficient. The most effective metric to assess the effectiveness of verification is the number of respins the SoC design takes to pass in the field. SA study on functional verification by Wilson Research group 2020 [1] claims that the percentage of new SoCs which have achieved first-time success is 32% which is alarmingly low. Added to this, the growing complexity of the SoC designs demands efficient verification methods to improve these statistics. This demand for effective verification for ASIC/SoC designs, which was seen during the middle of the year 2000, started devising many innovative methods to verify SoC designs, but there is still scope for more.

SoC verification is the process used to confirm the functional correctness of a SoC design. Aggressive time to market schedules and correct first-time requirements (SoC design working as intended when it is fabricated for the first time is called a first-time requirement or first-time success) exerts phenomenal pressure on the system verification at the design stage. A typical SoC design cycle, starting from specification to design tape out ranges between six months to three years depending on the technology, the complexity of the system, and the availability of the building blocks of SoC design. The fabrication process, packaging, ATE testing, and getting engineering samples for field validation (where chips are delivered to customers for product trials) typically take 6 months. Therefore, all, the SoCs are available for production only after the engineering samples are validated in the identified product use case scenario. If this is successful, SoC design is considered for mass production. This is the first-time the design has been successful (first-time success). Failures in any of these steps of the development cycle impact the design time exponentially,

© The Author(s), under exclusive license to Springer Nature Switzerland AG 2022   135
V. S. Chakravarthi, *A Practical Approach to VLSI System on Chip (SoC) Design*,
https://doi.org/10.1007/978-3-031-18363-8_7

sometimes requiring new metal tape outs with design corrections. Another driving factor for making the design work the first time is the fabrication cost of the nanometer technology. The typical fabrication cost of a 36 sq.mm. chip design in 40 nm CMOS FinFET technology is approximately 800000 to 1 million USD. High nonrecurring engineering (NRE) and fabrication costs incurred during the engineering sample development of SoC are to be amortised over the production in large numbers. So, if the NRE requires multiple tape-outs for the engineering samples, then it may impact business to such a large extent that it may not be viable at all commercially. Hence, first-time success is absolutely necessary in system on chip development. The feasibility of SoC design verification depends on identifying a set of "most common use case scenarios" of the system at the pre-silicon stage. This is a very complex and challenging phenomenon, as there can be innumerable use case scenarios. For example, one can easily imagine the innumerable use case scenarios of a smartphone mobile SoC with the primary function of a talking phone. But a smart mobile phone is used for many applications such as sending messages, shopping, tracking human health, banking, and infotainment. There will be a large number of application scenarios to test and validate the mobile SoC imagining of these application scenarios to identify and validate them. The cost of debugging the issue in SoC increases by a factor of 10 as the design progresses from one phase to the next in the development cycle. That is, the verification cost at the design phase is ten times less expensive than the verification of the same function at the wafer stage, which is ten times less expensive than verifying it at the chip stage. This is ten times less expensive than verifying it in the field at customer site. This is due to the higher debug access and the tool support the designer gets to the SoC design during the design than at advanced stages of development. Hence, a set of critical scenarios that are close to the actual applications' use cases are identified and targeted during the pre-silicon stage to get good confidence of first-time success of the SoC. Designing a SoC is done by integrating design blocks or IP cores of different types (soft, hard cores) which further challenges the design verification. The SoC design process also involves a number of design transformations from RTL to netlist and then to layout structures, which are then converted to mask data as shown in Fig. 7.1. When design goes through these transformations, it is required to verify that the design intent is retained at all levels untill it is fabricated. It is only SoC design verification that can guarantee it.

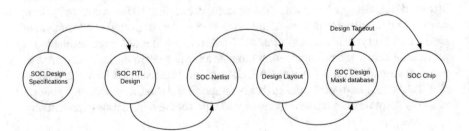

Fig. 7.1 Design transformations

To summarize, the reasons why verification is an important for SoC design are as follows:

- Exorbitant cost of fabrication demanding first-time success as multiple respins may make it commercially nonviable.
- Cost of verification increasing by a factor of 10 as the design progresses in development cycle. So early verification will boost confidence of getting the SoC design first time right.
- Since the SoC design involves series of transformations of database using EDA tools, it is essential to verify that these transformations are implemented correct which is done by verification.

7.2 Verification Plan and Strategies

For the first-time success (SoC working as intended when it is fabricated for the first time) of the VLSI SoC design, it is important to adopt many methods of verification at the design stage. Different types of verification techniques used are simulation-based verification, formal verification, timing verification, FPGA validation, and hardware emulation and validation. Verification by simulation was the only technique followed in the past. But with the growing complexity of systems, it is necessary to use every possible way to verify SoC designs.

It is difficult to define the condition for completion of design verification as it is almost impossible to simulate all the design scenarios of the SoC designs. Consider a design example of a single flip-flop which has two states; the number of test pattern required to test the flip-flop is 4. The ARM Cortex M4 core has 65 K gates in 65 nm technology and the gates can have multiple input-outputs. Just to simplify the discussion, assuming all gates have only two states, imagine the number of patterns required to test ARM cortex M4 core. It will be $65 \times 1000 \times 4 = 0.26$ million patterns. Just simulating all of them (without considering the problems of accessing them from primary input-outputs, finding the test patterns for each of them, etc.) using fastest of computer multiple times at different design stages is practically impossible (Fig. 7.2).

At the system level, also, identifying all the scenarios is very challenging. This could be because of the inability of the prediction and visualisation of the use case scenarios itself or may be due to the requirement of some more models or modules in the environment to realize it. It could be a full software stack, or a hardware platform on which the entire design database is ported, or a computational system infrastructure for the simulation. Hence, it is required to define, as the scope of pre-silicon verification, realizable scenarios as the verification test environment and a set of test cases. This can be approached in many ways:

- Top-down approach.
- Bottom-up approach.
- Platform level verification.
- System level or transaction level verification (TLV).

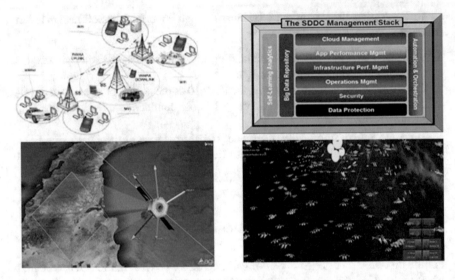

Fig. 7.2 Complex use case scenario of SoCs is difficult to model during design stage

Top-down level approach: In this approach, SoC is verified from the top most level of hierarchy for interfaces and then continued to the next lower level of hierarchy till the smallest leaf level design element is verified. Traditionally, this approach was used as a verification plan when the SoC design had a single or two levels of hierarchy.

Bottom-up approach: This is the most commonly used approach in design verification. It starts with the verification of smaller design blocks; verifying the small block is easy and practical. Also, finding bugs and fixing them is easier in block level simulations. This is because it is easier to trace the signals back and forth in the smaller design to debug an issue if one is found. As a number of blocks are verified, they are integrated to form the top module of the chip, which is verified by a separate top-level test setup. For example, the cores in a SoC consisting of a UART core, a USB core, and protocol bus interface cores are verified individually, core by core, first and then verified at the chip top level.

Platform level verification: If the design is based-on standards, like the USB device core, it is good to verify it on the standard platforms where a standard peer device like a USB host device is mounted. Similarly, a SPI slave core can be verified on the platform with a SPI master device. This will also confirm interoperability issues.

System interface-based or transaction level verification: If the SoC is protocol based, it is required to build the verification setup with a standard verification IP (intellectual property) core by monitoring the responses to the transactions. For example, the Wi-Fi device core is verified in an environment with the WLAN access point by observing the transactions between the two. WLAN access point core is a standard reference verification IP that is pre-verified and validated. This also proves the interoperability of the cores when fabricated.

7.3 Verification Plan

Verification plan is the document that describes the plan for SoC design verification and the tape out criteria. It explains how each of the functionalities of the SoC design is planned to be verified. It lists the verification goals at the module level and top level of hierarchy. It identifies the necessary tools, such as simulators, waveform viewers, and scripts used for verification. It explicitly mentions the coverage criteria for successful completion of verification as the finish criteria for design tape out. Different verification coverages relevant for SoC design are functional coverage, code coverage, and finite state machine (FSM) coverages. Functional coverage quantifies the number of functions to be verified by writing the test cases and setting the correct design response to be achieved by the simulations. There are tools that measure the functional coverage by going through the test cases and function (feature) lists. Since the functionality identified and fed into these tools is done manually, there is scope for underfeeding the number of functionalities to get the high percentage coverage. There is another parameter that is generally used called "code coverage" which determines the number of RTL statements covered by the test cases simulated on the design database at the RTL level. This helps to identify the redundant code in the design database and facilitates code cleanup. The tools used for code coverage are also capable of giving the finite state machine states covered by test cases. This is a very important measure used to cover all state transitions by adding appropriate test cases. The design verification coverage is used not only to assess the status of design verification in some companies but also to assess the performance of the verification engineers. The verification plan document lists criteria of completeness of design verification in terms of design coverage. The remaining gap in verification coverage is filled by other validation techniques like FPGA-based validation, emulation techniques, and testing SoC designs on the development boards. For example, if the functional coverage achieved by simulation is 98%, the remaining 2% is attained by porting the design onto an FPGA and testing the relevant functionality on the FPGA board or any other appropriate test techniques. These methods may require additional circuits on board and on FPGA to make it the test setup suitable for functional validation of the SoC design. It may also require software running on board or system interfaced with it. Typical test setups used for SoC verification in a simulation environment, an FPGA environment, and on a development on the board are shown in Fig. 7.3.

The major design details contained in verification plan are the following:

1. Definition of first-time success for the SoC design.
2. Critical application scenarios of the SoC. The requirement for the development of test environment for SoC testing.

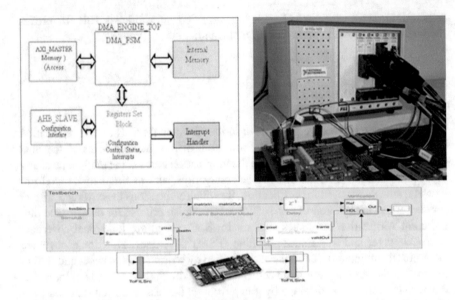

Fig. 7.3 Test benches to simulate use case scenario of VLSI SoCs

3. Development plan for the functional verification environment and resources required.
4. List of functional features to be verified at module level and at the top levels of design hierarchy.
5. Main verification strategy for both blocks and top-level design.
6. Test bench modules at design RTL level:

   (a) Bus functional module (BFM) and bus monitors.
   (b) Signal monitors.
   (c) Verification reference models.

7. FPGA level validation details:

   (a) Requirements of FPGA board needed for SoC validation.
   (b) Additional module needed for FPGA validation platform.
   (c) Software modules required, software development, and debug platforms will be developed based on this requirement.

8. Required verification tools and flows.
9. Requirement for block level simulation environment.
10. Regression test environment and plan of regression testing.
11. Clear criteria to determine the completion of verification such as target coverage, number of regression test vectors, and gate level simulation strategies and expectations.

Design resources include verification engineers with their skill set, hardware development boards, FPGA boards, software requirements, an EDA tool environment,(workstations and servers) simulators (number of licenses), and the design infrastructure for verification. The strategy to verify the VLSI SoC varies with the design complexity and the use case scenario of the SoC. Ideally, it is targeted to emulate/simulate the use case scenarios using the test bench at the RTL level or the FPGA verification plan or using the development board setup, or a combination of any or all of them. Using these resources, the SoC design is verified to gain a high level of confidence to predicting its success. The verification strategy involves design partitioning for verification at the subblock level and chip top validation on FPGA boards.

7.4   Functional Verification

The goal of functional verification is to confirm that the SoC design functions as intended in the functional scenarios as well as in its application scenarios. One use case scenario can be mapped to one or many functional test scenarios. For example, to verify the addition function of the block, there could be three test cases: the first one that verifies the input operands the second one that verifies the output results corresponding to the inputs, and the third one to check the carry operation of the adder. Basically, SoC design contains multiple blocks of different functionalities, interconnected with each other and/or a shared bus on which a number of blocks interact, or a block functioning as per the standard protocol. In such cases, functional verification of a SoC involves simulations of (a) block-to-block interface verification, (b) bus contention verification, and (c) protocol and compliance verification.

7.5   Verification Methods

There are three types of design verification. They are black box, white box, and gray box verification. SoC design is verified by adopting different combinations of these methods.

7.5.1   Black Box Verification

This is a verification method where the internal details of the design implementation are not exposed to the verification. Verification is done by only accessing the exposed interface signals without accessing internal states or signals, thereby making it implementation independent. Obviously, the verification will not be visible to the design's internal implementation details or system states. This method is best suited to uncover interpretation level issues like endianness checks, protocol misinterpretations, and interoperability tests.

7.5.2   White Box Verification

In this verification method, the test bench modules can access internal states, signals, and interfaces of the design. It is very easy to debug any design issue in this because the test bench can literally trace the signal drivers with the expected in mind. This method is best suited for checking low-level implementation-specific scenarios and design corners where they can target the design for the scenario that has potential issues and debug them. An example of such scenario is FIFO pointer roleovers, counter overflows, etc. Assertions are best suited for checking internal design behaviours in this method. This method is totally complementary to the black box verification method.

7.5.3   Gray Box Verification

This method is intermediate between black box and white box verification techniques. In this method, the test environment verifies the system at the interface levels with IOs at the top levels with and on need basic (like for design corners) access design internals for test and debug. Typically, first-level tests are targeted using the black box method, and the functional coverage is assessed. To improve the coverage, if required, through a white box approach, the test scenarios are tested.

7.6   Design for Verification

With SoC design methods moving toward the system or architectural level, it is essential to verify the system functions at the transaction level across subsystems. However, SoC design is predominantly integration of predesigned or pre-verified IP cores, which is more like black box verification for the internal IPs. Also, the complex SoC design is tending toward being verification friendly where the internal states and critical signals are latched and made available for software to read through the primary interfaces and hence predict the root cause of the issue. This will be useful in "black box" or "gray box" verification. Functional verification is done differently in different environments. In the RTL level, test bench and a set of test cases are developed and simulated using the simulators to see if the SoC behaves as intended. The functional correctness is checked by viewing the waveforms at the interfaces or module/block level inputs and outputs.

In the FPGA-based hardware validation, the design under test in RTL form is ported to the FPGA on the board, limited software is run, actual stimulus is fed to the SoC input, and output is observed on the development environment.

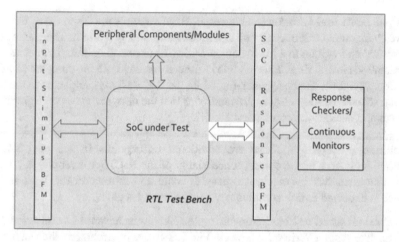

Fig. 7.4 RTL Test bench internal modules to simulate use case scenario of VLSI SoCs

On the development environment, the development platform based on the sub-modules is designed and developed with interfaces as close as the final SoC and is verified with some more complex software.

Test environment or test bench at RTL level represents the most likely environment in which the SoC is intended to be used. All the environments are developed to accept the stimulus as close to the real-world inputs as possible. A typical RTL test environment (also called test bench) is shown in Fig. 7.4. It is a closed system as it represents a complete environment, including the input stimulus and output controls through behavioural bus functional models (BFM).

SoC Under Test It is the SoC design whose functional intent has to be verified.

Peripheral Modules These modules are support modules which are required to make the SoC under verification complete in the application environment. They are the verification IPs or IPs of peripheral functions, like external memories, data converters and real-time sensor models.

Input Stimulus and Bus Functional Model (BFM) The input stimuli represent the input signals that the SoC under verification is fed with from the external world in the real application scenario. It can be system design signals like the clock from the reference crystal, the reset signal, sensor inputs, or data inputs from modules or verification IPs that are external to SoC. Generation of the stimulus from different source as required by the SoC is automatic (when the reference clock is fed to the PLL module, it automatically generates a system clock of the required frequency for the SoC as configured) or semiautomatic with a manual trigger or conditional. They are fed to the SoC design through the interfaces, following the timing requirements of the design through the bus functional model (BFM).

Output BFM and Checkers This output BFM captures the response of the SoC through its output interfaces when a particular stimulus is fed to it. The response is compared and written to a file to compare with the expected outputs, or checked for expectations in real time. This module is either a checker with file compare capability or a waveform database generator, while the SoC design is subjected to the particular scenario through test input conditions that the designer views on a graphical viewer and decides the correctness of.

Continuous Monitors These are additional checkpoints in the environment that are indicators of the correct functionality of the SoC. For example, in a timer SoC that generates 1-s clock, it is easy to continuously monitor the 1-ms signal, which is expected to tick continuously to generate a 1-s clock.

More advanced test environment implementable in advanced verification languages like system Verilog is shown in Fig. 7.5. In test environment, the test blocks are very modular and the results are automatically checked, and a pass/fail decision is taken, hence they are automation friendly. The test environment is capable of

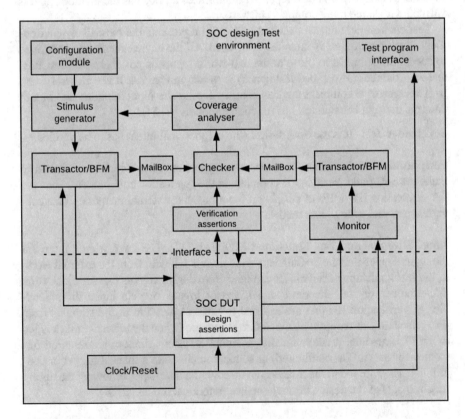

Fig. 7.5 Automated test environment

analyzing the design for functional, code, and FSM coverage. A brief description of the modules of test environment follows.

SoC DUT: The SoC DUT is the SoC design under test which is to be verified.

Design and Verification Assertions The design under test and the verification test environment can have assertions to improve the effectiveness of verification. Assertions are the statements that are used to check the temporal relationship of synchronous signals in the design for correct functioning of the module. The design assertions, if supported, are tracked by the test bench checker module to see if it has triggered or not and is assessed for correctness. For example, consider a part of the logic design where a functionality is to check if the received packet is correct, and the packet received is validated by the packet_valid signal. It is obvious that the packet_valid signal should be set high whenever the packet_correct or packet_error signal is generated. In this context, it makes sense to write a design assertion that checks the co-occurrence of packet_error and a packet_valid or packet_correct and packet valid signal, and if the assertion gets triggered, the design intent can be verified. In the example shown, a design assertion is written to see if packet_valid and packet_correct or packet_valid and packet_error signals don't co-occur. If this assertion is triggered, the design is faulty. This is shown in the timing diagram in Fig. 7.6.

Similar assertions can be written at the transaction level of DUT transactions, which are tracked for the correctness of the design.

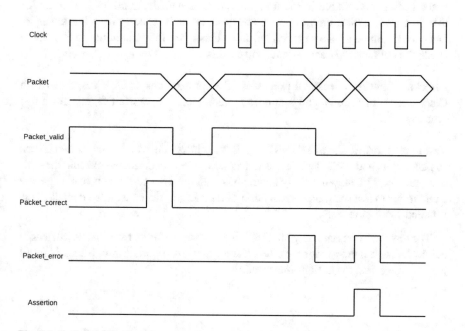

Fig. 7.6 SoC design logic with an assertion

Clock/Reset Block   Clock reset block generates the required clock and reset signal as per the requirements of the SoC design.

Configuration   This block sets the DUT and test bench in the desired configuration in which the DUT has to be tested.

Stimulus Generator   This module generates the input stimulus in the test bench. Typically, this module generates signals in the required order and sequence as per SoC functionality. It can be a complex IP verification.

Transactor/Bus Functional Module (BFM)   The transactor or bus functional module follows the interface specification to feed the stimulus to the SoC DUT. There will be as many BFMs as there are bus interfaces. If the SoC design supports UART, USB, and PCI Express interfaces, there should be a BFMs corresponding to each of these interfaces that manages transaction compliance with these protocols.

Mailboxes   These are communication mechanisms in system Verilog test bench that allow messages to be exchanged between processes. The process that wants to talk to another process posts the message to mailbox, which stores the messages temporarily in a system-defined memory object, to pass it to the desired process. Mailboxes are created with either a bound or unbound queue size. A bound mailbox becomes full when it contains the maximum number of messages defined. A process that attempts to place a message into a full mailbox shall be suspended until enough space becomes available in the mailbox queue. Basically, mailbox is a technique that synchronize different processes. The process can be checker as in this example. Once the mailboxes have a predefined set of messages, they can initiate a checker to check the content and decide on its correctness.

Checker   Checker checks all processes, like comparisons of DUT responses with expectations, assertions, and monitors to decide on the pass/fail criteria for a test scenario.

Test Program Interface (TPI)   This is the user interface, which accepts user inputs as parameters and compiles options to trigger the test scenario and execute the simulations. The TPI supports many commands with optional parameters to execute the simulations in test scenarios, one by one, and generates the consolidated result. This is called regression tests.

The test environment shown in Fig. 7.5 can be extended to the most user-friendly automated test bench, which can even send the test reports through emails to all parties concerned to get their intervention.

7.7   Verification Example

In this section, the simulation of a simple decade counter design is presented for a clear understanding of the verification process.

Design functionality of the decade counter: The decade counter counts numbers 0, 1, 2, 3, 4, 5, 6, 7, 8, 9, 0 at every active edge of clock when it is enabled. It is a design requirement that an output signal be generated whenever the counter counts 5. The pin diagram and test bench of the decade counter are shown in Fig. 7.7.

The Verilog module and the test bench model of the decade counter is shown in Fig. 7.8.

The test bench module of the decade counter is shown in Fig. 7.9.

The design file is saved as decade-counter.v, and the test bench file is saved as tb_dcounter.v (.v represents the Verilog file) in the present working directory. To simulate the design file, a simulator is used. Most used simulators are cycle-based simulators. Cycle-based simulators sample the signals and compute the design response every clock cycle. The simulator first analyzes the RTL code and elaborates before simulating the design.

As the simulation is executed, observe for log messages displayed on the terminal for errors and warnings. If there are any errors/warnings, it is required to correct them in the design files. For the modules in the design example, there should not be any warning or error and simulation terminates with success. If you observe in the present working directory, there are many output files generated by the simulation run. They are command log file, waveform dump file named decade_counter.vcd. The decade_counter.vcd file can be opened with the waveform viewer tools. When this file is opened in the waveform viewer tool, one can observe the logic state changes on the input-output signals and internal nets. For more information on running the simulations and using the waveform viewer tools, one can refer to the respective user manuals for help. The design behavior is verified by observing design signals, clock, reset_n, and out_5, count_out. The waveform looks like the one in Fig. 7.10.

The verification flow can be extended to the designs of any complexity. The next design example explained in this section demonstrates this. The verification of self-synchronizing descrambler which uses scrambler design as verification IP in the test bench is explained. Consider the design of a self-synchronizing scrambler with

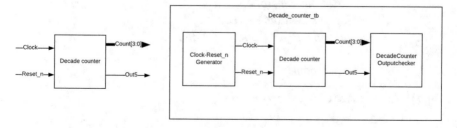

Fig. 7.7 Decade counter as design under test and decade counter test bench

```
module decade-counter {
                        clock,
                        reset_n,
                        counter_enable,
                        out_5,
                        countupto9
                        };

//input -output declaration

input        clock, reset_n, counter_enable;
output       out_5;
output [3:0] countupto9;

//Internal signal declaration

reg [3:0]    counter;

wire         out_5;
wire [3:0]   countupto9;

//Logic description

always @(posedge clock or negedge reset_n)
begin
  if (reset_n == 1'b0)
     counter <= 4'd0;
  else if (counter_enable == 1'b1)
      if counter < 10)
         counter <= counter +1;
      else counter <= counter;
  else counter <= 4'd0;
end

//Output generation
assign countupto9 = counter;
assign out_5 = (counter==4'd5);

endmodule
```

Fig. 7.8 Verilog module of the decade counter design

```
module tb-dcounter;

//Internal signal declaration

reg     clock;
reg.    reset_n;
reg.    enable;

initial
begin
  #10 clock    <= 1'b0;
  #10 reset_n <= 1'b0;
  #10 reset_n<= 1'b1;
  #10 reset_n<= 1'b0;
       Enable <= 1'b0;
End
Initial #50 enable <= 1'b1;

always @
 #10 clock <= ~clock;

//Module instantiation
decade-counter udecade-counter {

                    .clock.          (clock),
                    .reset_n.        (reset_n),
                    .counter_enable (enable),
                    .out_5.          (out_5),
                    .countupto9.     (count_out)
                    };




//display commands for waveform generation
 Initial
begin
 $dumpfile(decade_counter.vcd);
 $dumpvars(1, tb_dcounter);
end

endmodule
```

Fig. 7.9 Test bench module for decade counter

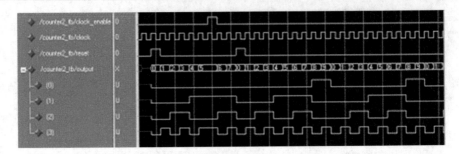

Fig. 7.10 Simulation waveform of decade counter

Side-stream scrambler employed by the MASTER PHY

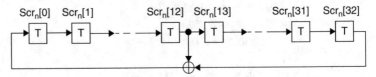

Side-stream scrambler employed by the SLAVE PHY

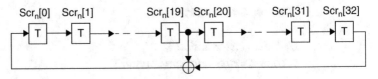

Fig. 7.11 Implementation of self-synchronizing scrambler

the polynomial $g(x) = 1 + x^{13} + x^{33}$. A self-synchronizing scrambler module is used in communication to scramble the incoming data if it is a long sequence of zeros or ones to have zero DC bias. The data is scrambled at the transmitter using the same polynomial and descrambled to recover the original data transmitted at the receiver end using the same polynomial. The functional property of the descrambler in the self-synchronizing descrambler in that it does not need to be initialized by the initialization vector to achieve synchronization. The implementation is shown in Fig. 7.11. Synchronization of scrambler-de-scrambler is defined as both the LFSRs of scrambler and descrambler holding the same pattern, and hence, when the data is fed to the descrambler, it can generate the input of the scrambler data.

The Verilog models of the scrambler and descrambler are shown in Figs. 7.12 and 7.13, respectively. The test bench file is shown in Fig. 7.14. The module under test is descrambler. To test if the descrambler synchronizes to the scrambler, it is required to have the descrambler LFSR reset to any initialization values. The random pattern is fed through the scrambler, and the scrambled data is fed as input stimulus to the descrambler. It is to be verified that the descrambler, at some point in time will be able to decode the incoming data. One may notice that the test bench will not have any ports, as this will be a self-contained environment for the module under test.

```
//-----------------------------------------------------
// 802.11B Scrambler
// Module Name : scrambler
// File Name : scrambler.v
// Function : This is a 7 bit scrambler for 802.11b
// Asynchronous and active high enable and active low reset signal
//-----------------------------------------------------
module scrambler (
clock ,    // Clock input of the design
resetn ,   // active low, asynchronous Reset input
enable ,   // Active high enable signal
bit_in,    // Input data bit.
bit_out    // Scrambled output bit.
); // End of port list
//-------------Input Ports----------------------------

input clock ;
input resetn ;
input enable ;
input bit_in;
//-------------Output Ports--------------------------
output      bit_out;

//-------------Input ports Data Type------------------
// By rule all the input ports should be wires
wire  clock ;
wire  resetn ;
wire  enable ;
//-------------Output Ports Data Type-----------------
// Output port can be a storage element (reg) or a wire
reg [6:0]    state_out ;
wire  bit_out;
//------------Code Starts Here------------------------
assign feedback = (bit_in ^ state_out[6] ^ state_out[3]);
assign bit_out = feedback;

// We trigger the below block with respect to positive edge of the
clock.
always @ (negedge resetn or posedge clock)
begin : SCRAMBLER // Block Name
if (resetn == 1'b0) begin
state_out <= #1 7'b1010101; //Striped start.
end
```

Fig. 7.12 Verilog model of scrambler module

```
// If enable is active, then we tick the state.
else if (enable == 1'b1) begin
state_out <= {state_out[5:0], feedback};
end
end // block: SCRAMBLER
endmodule
```

Fig. 7.12 (continued)

The test bench consists of the following sections:

- The first section in the test bench will be the stimulus generation which includes clock, reset, enable, and data generation.
- The second section is the scrambler block, which is used as standard verification IP.
- The third section is the module instantiation.
- The fourth section is the output reader and waveform dumping for debugging and user verification.

The test bench sections are shown in Fig. 7.15. A typical SoC test bench will have multiple clock (OCC) generation blocks with standard PLLs, multiple VIPs as needed, and control state machines that will enable each of these modules for multiple test scenarios. The output reader and waveform dump section can be complex blocks that can automatically verify the correctness of the functionality depending on the SoC verification requirements.

More simulation examples can be found in Chap. 11 "Reference Design." Reader can actually simulate the designs and verify the results to compare with sample waveforms to check the correctness.

7.8   Verification Tools

There are a number of verification tools which are used for the functional verification of SoC design. They are the following:

- Simulators.
- Coverage tools.
- Lint tools.

Among the above listed tools, simulators are indispensable for RTL functional verification. A simulator is the tool that is executed to understand the design behaviour in most anticipated use case scenarios by using test vectors in a test bench. It is a software that enables the study of SoC design states and its outputs in the presence of user-fed stimulus for the required duration, called the test vectors. There are different types of simulators. They are cycle-based simulators, event-based simulators, and circuit simulators. The SoC design to be simulated is called the device under

```
//------------------------------------------------------
// 802.11B Decrambler
// Module Name : descrambler.v
// File Name : descrambler.v
// Function : This is a self-synching 7 bit descrambler for 802.11b
// asynchronous active low reset and with active high enable signal
//------------------------------------------------------
module descrambler (
                clock ,   // Clock input of the design
                resetn ,  // active low, asynchronous Reset input
                enable ,  // Active high enable signal
                bit_in,   // Input data bit.
                bit_out   // Scrambled output bit.
                ); // End of port list
        //-------------Input Ports---------------------------
        input clock ;
        input resetn ;
        input enable ;
        input bit_in;

        //-------------Output Ports---------------------------
        output    bit_out;

        //-------------Input ports Data Type------------------
        // By rule all the input ports should be wires
        wire       clock,resetn,enable ;
        //-------------Output Ports Data Type-----------------
        // Output port can be a storage element (reg) or a wire
        reg [6:0] state_out ;
        reg       bit_out;

        //------------Code Starts Here------------------------
        assign feedback = (bit_in ^ state_out[6] ^ state_out[3]);

        // We trigger the below block with respect to positive edge of the
        clock.
        always @ (negedge resetn or posedge clock)
          begin : DESCRAMBLER // Block Name
            if (resetn == 1'b0) begin
                //Self synching, so  reset value can be anything. Only for
        simulation.
```

Fig. 7.13 Descrambler Verilog module

```
     //This might cause a problem in synthesis.
       state_out <= #1 7'bXXXXXXX;
     end
     // If enable is active, then we tick the state.
     else if (enable == 1'b1) begin
       state_out <= {state_out[5:0],bit_in};
       bit_out <= feedback;
     end
   end // block: DESCRAMBLER
endmodule
```

Fig. 7.13 (continued)

test. The simulator, using certain commands in the test bench, can monitor and write out the internal logic levels, states of the signals in the design module and input-outputs during the simulation. This waveform output file is then opened in the waveform viewer tools that interfaces with the graphic debug environment. Different simulators are used for verification based on the type of SoC design. Cycle-based and event-based simulators are digital simulators. Most of the simulators used for digital simulations are cycle-based simulators. Cycle-based simulators evaluate the design for its logic states every cycle. Simulator cycles are of the order of pico or nanoseconds to virtually emulate the concurrent behaviour of hardware for the user. Above mentioned simulators are all cycle-based simulators. They are called cycle-accurate simulators because they sample the SoC design at the input edges of clock signal. An example of timing waveform from cycle-based simulators is shown in Fig. 7.16. The cycle-based simulators are 10–100 times faster than the event-based simulators and are used in majority of the SoC design verification. Design verification, which uses cycle-based simulators, requires STA analysis as the design is verified at clock intervals.

Event-based simulators evaluate the design whenever a logic change happens on any of the nets in the circuit. These simulators are also called timing-accurate simulators and are suitable for small circuit level verification. They provide a good debug environment and also do not require timing analysis as the design is functionally verified at all the events on all the nodes in the design. Event-based simulators require large computing machines on which the simulation is run. This is because of the explosive number of nets in today's SoC designs, which will have large logic transitions during simulations. Monitoring the large number of logic transitions on the nets and evaluating them in all their combinations is practically impossible. Debugging the fault in such a design is very difficult. An example of timing waveform of design simulated by an event-based simulator is shown in Fig. 7.17.

Typical tool flow in event-based simulator engine is shown in Fig. 7.18.

```
`timescale 1ns/10ps

//Module definition and signal declaration
module tb_top;

reg Clk;
reg Resetn;
reg Enb;
reg [7:0] Pattern;
reg [7:0] DataIn;
reg [7:0] DataOut;
integer  errCnt;
integer  CompFlag;

reg     Match;
wire    Din;
wire    Sout;
wire    Dout;

//clock generation
  always #5 Clk = ~Clk;

assign Din = DataIn[7];

// Application of Stimulus
initial
begin
  Clk = 0;
  Resetn = 0;
  Enb =   0;
  CompFlag =0;
  errCnt = 0;
  Match = 0;
  $display("--------- Test  Started ---------");
  #100ns Resetn = 1;

  $display("--------- Sending Data Patternn : 0x55 ---------");
  repeat (10)  @ (posedge Clk);
  Enb = 1;
  Pattern = 8'h55;
  DataIn   = Pattern;
  repeat (100) begin
    @ (posedge Clk) #1 DataIn  = {DataIn[6:0],DataIn[7]};
  end
  repeat (10)  @ (posedge Clk)Enb = 0;
```

Fig. 7.14 Test bench file where the scrambler-descrambler modules are instantiated

```
$display("--------- Sending Data Patternn : 0x11 ---------");
repeat (10)  @ (posedge Clk);
Enb = 1;

Pattern   = 8'h11;
DataIn    = Pattern;
repeat (100) begin
    @ (posedge Clk) #1 DataIn  = {DataIn[6:0],DataIn[7]};
end
repeat (10)  @ (posedge Clk)Enb = 0;
CompFlag  = 0;

$display("--------- Sending Data Patternn : 0x22 ---------");
repeat (10)  @ (posedge Clk);
Enb = 1;
Pattern   = 8'h22;
DataIn    = Pattern;
repeat (100) begin
    @ (posedge Clk) #1 DataIn  = {DataIn[6:0],DataIn[7]};
end
repeat (10)  @ (posedge Clk)Enb = 0;
CompFlag  = 0;

$display("--------- Sending Data Patternn : 0x33 ---------");
repeat (10)  @ (posedge Clk);
Enb = 1;                                //Application of Stimulus
Pattern   = 8'h33;
DataIn    = Pattern;
repeat (100) begin
    @ (posedge Clk) #1 DataIn  = {DataIn[6:0],DataIn[7]};
end
repeat (10)  @ (posedge Clk)Enb = 0;
CompFlag  = 0;

$display("--------- Sending Data Patternn : 0x44 ---------");
repeat (10)  @ (posedge Clk);
Enb = 1;
Pattern   = 8'h44;

DataIn    = Pattern;
repeat (100) begin
    @ (posedge Clk) #1 DataIn  = {DataIn[6:0],DataIn[7]};
end
repeat (10)  @ (posedge Clk)Enb = 0;
CompFlag  = 0;
```

Fig. 7.14 (continued)

```verilog
    $display("--------- Test Ended ---------");
    #10  $finish;
end

always@(posedge Clk)
begin
  if(Enb) begin
    DataOut = {DataOut[6:0],Dout};

    #1 if(DataOut == Pattern) Match = 1;
      else Match = 0;        //Application of Stimulus
  end
  else DataOut = 8'hXX;
end

//SoC module instantiations
scrambler u_scarmb(
            .clock   (Clk),   // Clock input of the design
            .resetn  (Resetn), // active low, synchronous Reset input
            .enable  (Enb),   // Active high enable signal
            .bit_in  (Din) ,  // Input data bit.
            .bit_out (Sout)   // Scrambled output bit.
            ); // End of port list

descrambler  u_descramb(
            .clock   (Clk),   // Clock input of the design
            .resetn  (Resetn), // active low, synchronous Reset input
            .enable  (Enb),   // Active high enable signal

            .bit_in  (Sout),  // Input data bit.
            .bit_out (Dout)   // De-Scrambled output bit.
            ); // End of port list

  initial
  begin
    $dumpfile("tb_top.vcd");
    $dumpvars(0,tb_top);
  end

endmodule
```

Fig. 7.14 (continued)

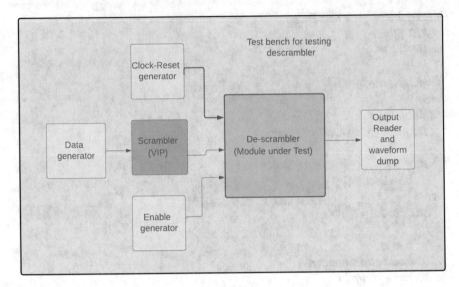

Fig. 7.15 Descrambler test bench block diagram

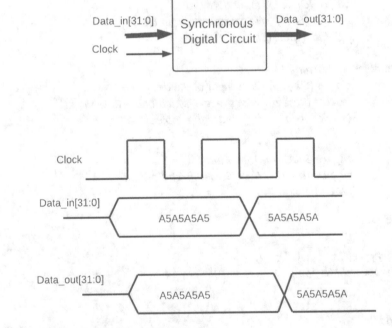

Fig. 7.16 Cycle-based simulator of the design

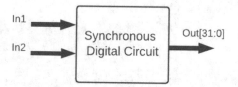

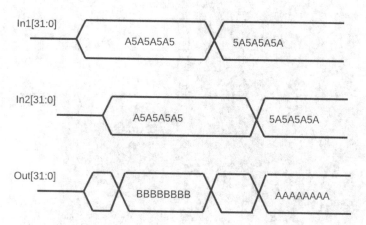

Fig. 7.17 Event-based simulation example

Fig. 7.18 Tool flow diagram in even-based simulations

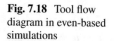

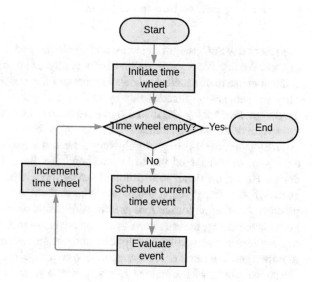

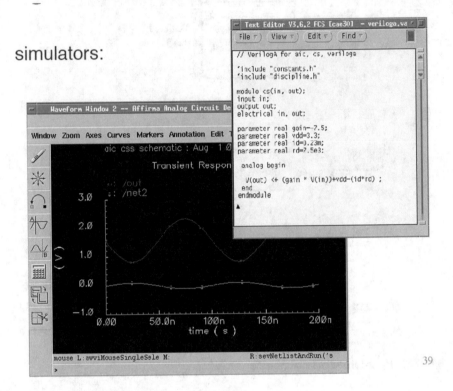

Fig. 7.19 Analog simulator snapshot of a design

Present day SoC designs, includes analog blocks, and it is required to verify them as well. Analog blocks are verified individually using analog simulators. Analog simulators use mathematical models to represent the analog functions of the design. They emulate analog functionality by sensing and generating suitable responses to the design. Few analog and mixed signal simulators are available. Figure 7.19 shows a snapshot of the analog simulator response for a design.

Analog simulators are generally very slow and are not automated. They require a designer to understand the design well and use the tool as an aid to analyze the design. Hence, the detail verification of analog modules is done separately, and then analog-digital mixed signal simulation is carried out just to verify the integration in practice. Another important tool used in verification process or module in the simulation is extracting coverage analyser. The coverage matrix gives insight to quality or completeness of verification done on the design database. There are three types of coverages: the functional coverage, code coverage, and state machine coverage. Functional coverage is obtained by comparing and analyzing the test cases run on the SoC design database and functionality feature checklist of the SoC design. Code coverage is the metric that is extracted when simulation is run on the SoC design to track the code lines in the design that are excited. The state machine coverages give information on state transitions in design FSMs due to the test case in the simulation run. All these matrices help the verification engineers maximize the coverage metrics, and hence reach the design verification goals.

Lint tools check the SoC design at the RTL level against the rules set for different objectives apart from the basic syntax and semantics of the HDL language. It is a static RTL code checker. It checks by compiling the design and preprocessing it for simulation, synthesis, and DFT simulations. Different design objectives where lint is run are the basic compilation of the RTL design for simulation, synthesizability, and testability. There are standard rules defined by the tool for each of these objectives. Each of these rule sets can be customizable or enhanced for SoC-specific design goals. When executed on the design files, the tools write out log files with detailed analysis of the design against the rules defined and alerts with warnings and errors on the violations depending on the severity of the violations. nLint and HAL are two of the few known verification tools used in design centers.

7.9 Verification Language

The languages used to model the test bench or test cases are more relaxed and flexible compared to design languages. The main reason for this flexibility is the need to create more randomness in the test cases, and this need not be synthesizable. Verilog, being one of the oldest HDLs is also the verification language. Owing to the change in design description methodology raised to higher abstraction levels at the architectural level, a few of the verification languages like SystemVerilog, Vera, and System C are emerging as major Hardware Verification Languages (HVLs) at higher abstraction layers. These languages support class, object oriented, class extensions, and temporal properties, which help define system level or transaction level test functions easily. Of the mentioned languages, SystemVerilog is also gaining popularity as a powerful assertion language, which is a major feature in verification. But it also provides constructs designed to ensure consistent results between synthesis and simulation. Also, there are simulation tools that support these language constructs and can interpret the results and analyze them in terms of test coverage. They support interfaces like direct programming interface (DPI) to high-level software languages like C++ and Java, which enable the building of graphical user interfaces (GUI) which can make the verification environment more generic and effective at higher levels of abstractions up to the system level of hierarchy. More details of these can be found in the language books mentioned in the references. The simulator tools are now intelligent enough to ignore the common mistakes made by the designers and have the option of self-correcting them by notifying the user of the warnings.

7.10   Automation Scripts

Creating use case scenarios for the SoC is achieved by a set of complex test cases with random stimuli as the real time scenario is random. When the stimulus is random, the response to such a stimulus becomes hard to predict. So, the tests are typically carried out in such cases by predicting the end results or status or analyzing the statistics and stability of the system and also at an intermediate state of the system that is predictable. This requires some sort of coherence with the input randomness and the system states, which are more or less changing. To map out this correspondence, the test cases are automated. Automation means the test expectations, like data integrity, status, and handing over control to the next random scenario, are automatically controlled and evaluated. This is achieved through scripting languages. Most used scripting languages include Perl, Tcl, PHP, etc. Hence, the scripting languages are programming languages written for special run time environments that automate the execution of tasks that otherwise could be executed by the user one by one. These constructs are also understood by the EDA tools and hence can be integrated in the test setup. Automation is also done for the analysis of large data for integrity checks, statistical analysis, and running the test case in batches to get the desired functional coverage. Test scripts are interpreted, and not compiled.

7.11   Design for Verification

Design verification guarantees the quality of designs by uncovering the potential errors in system design and architecture. This is possible only when all functions of the system are simulated as exhaustively as possible while carefully investigating any possible erroneous behavior. This deserves the most time, attention, and complete knowledge of design use cases. This, in most cases may become very challenging if design is complex. This demands that the design be verifiable. It is the designer who has a complete understanding of the design implementation of a functionality. If the designer identifies critical design corners and critical states of the design, verification can be targeted to monitor and check the same. Sometimes in designs, certain scenarios may require long simulation runs to hit the design corners, which the verification engineer may not know. A simple example is the overflow generation of 32-bit counters, which run on a one second clock. This takes a long time to simulate but may happen quickly in real hardware. In such cases, if the designer provides a feature to preload a counter, 32-bit counter overflow is feasible. Such design tweaks make designs verifiable for more scenarios. The designer must identify critical design corners that can be addressed for verification. In addition, nonfunctional features of the system like scalability, expansion, and flexibility require extra design support to target them for verification. Such examples are memory address expansion, access by software writing to registers or memories in nondefault mode, extra configurations provided as an alternative to probable misinterpretation issues such as little endian or big endian, etc.

7.12   Assertions in Verification

Implementing assertions within a design requires a conscious decision to view the design process differently. This is an additional design and verification statement that is used to monitor this part of the design for its correctness. The assertion will definitely reduce the debugging time and effort. Assertions essentially act as early warnings during simulation can pinpoint failures that may either directly cause test failures or not be detected by passing tests. Assertions on module interfaces can quickly identify invalid behaviour that may be caused by a behavioural model or improper use of the design (invalid register settings, invalid operating modes, etc.). Such assertion failures indicate that a problem may be with the test bench, which helps the verification engineer fix issues in the test bench. It helps to fix issues in the design for its misbehavior. Design assertions help locate the root cause of failures by looking at the incorrect functions shown by them. For example, the constraint random simulations detect design issues at design corners at overflows and under-flows of FIFO operations that are not typically targeted by the directed tests. Simple assertions on FIFO interface signals detecting simultaneous read and write operation, number of reads exceeding the number of writes, etc. will help to find the root cause of actual failures during the test scenario without lengthy debug sessions. The bigger advantage of assertion is that it makes the design and verification test benches reusable by preserving the design intent beyond the design and verification owners.

7.13   Verification Reuse and Verification IPs

Just like design blocks that are reusable, verification modules can also be reused across generations of SoC designs. With multiple interface protocol blocks being part of the SoC, like a few SoCs that have multiple USB cores, multiple SPI cores, and multiple UART cores, the corresponding test modules can be reused in the test benches. Bus interface modules (BFMs), and interface cores in test benches can even be used to verify the number of SoCs that has the same functionality. This will also address time to market reduction and the design productivity gap. With SoC function becoming more and more complex, with many integrated cores complying with many standards and required to be interoperable, it has been the practice in the past couple of decades that the modules are developed as reference models, ensuring compliance with standard specifications. These are called verification IPs. These are pre-verified or certified for compliance with standard or protocol specifications. These can be licensed, or purchased on royalty terms from the IP developers. These VIPs are integrated as standard IPs in the test environments and a SoC are tested against verification IP to prove compliance and interoperability. Reuse of verification IPs is a common practice in SoC verification.

7.14   Universal Verification Methodology (UVM)

Universal Verification Methodology (UVM) is an industry standard verification methodology to define, reuse, and improve the verification environment and to reduce the cost of verification. It provides certain application programming interfaces (APIs) for the use of base class library (BSL) components in the verification environment making them reusable and tool independent. UVM-based verification environment is flexible enough for various types of test creation, coverage analysis, and reuse. The UVM standardization has improved interoperability and reduced the cost of repurchasing, and rewriting intellectual property (IP) for each new SoC design or verification tool is adopted to make it easier to reuse verification components. Overall, the UVM standardization will lower verification costs and improve design quality throughout the industry. More importantly, it can be implemented using SystemVerilog which is the most commonly used in complex SoC design verification.

7.15   Bug and Debug

Bugs are defects in the system. The quality of the SoC designs is directly dependent on the defects or bugs hidden in it. As stated earlier, the cost of testing at a higher design or development phase (RTL, physical design, layout, chip, board, system, system in field) is atleast ten times higher than the cost of testing at lower design or development phase. It is wise to uncover the defects or bugs at the earlier design/development phases. Bug is the unwanted states or conditions for the particular scenario. It can be temporary or permanent. This can arise for many reasons. The predominant reason would be the inability of the designer to interpret the requirement as desired (refer to the famous tree swing example in the figure on the requirement-interpretation issue) and due to lots of implicit, unstated requirements. Design bugs can also seep through because of the interpretation of system requirements by the verification person and his ability to create test cases for the entire use case scenario. It can also be because of human error and the tool errors that are used to do the design transformations during the design stage. During the design and development stage or in the field, it is essential to formally log and manage the bug so that it is fixed and does not appear again and again (Fig. 7.20).

7.16   Bug Tracking Workflow

Formal bug tracking is very essential in the design/development cycle to make sure the bug is understood correctly and is closed with the design fix if required. Due to the complexity of systems and the multiple design and development teams working

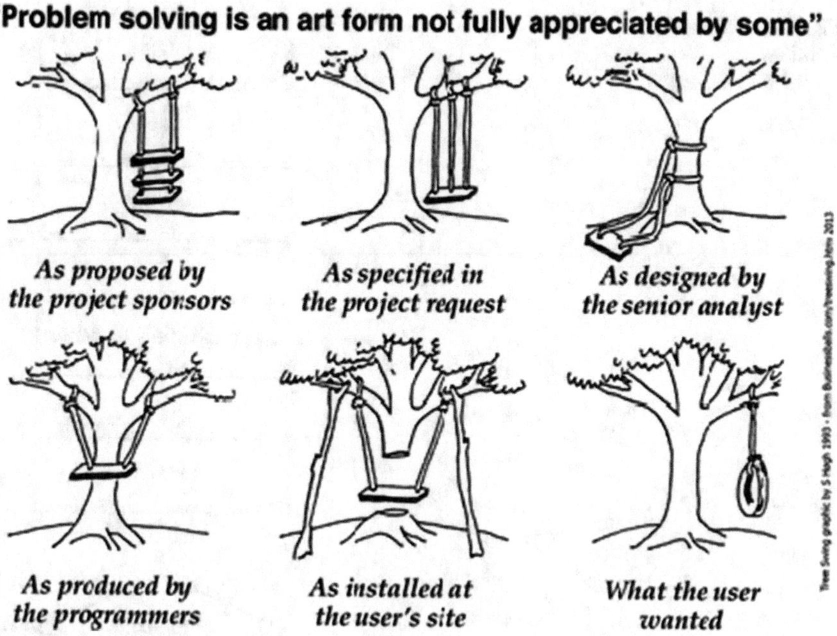

"Problem solving is an art form not fully appreciated by some"

**As proposed by
the project sponsors**

**As specified in
the project request**

**As designed by
the senior analyst**

**As produced by
the programmers**

**As installed at
the user's site**

**What the user
wanted**

Fig. 7.20 Tree swing example demonstrating the interpretation issues of requirement and the departmental barriers

on it, bug tracking tools are used. Tools enable formal tracking of the bug resolution. The bug tracking tool supports reporting the issue (logging), assigning to design owners, tracking the status of the fix, and solving and confirming that the issue is resolved by reverification. Stryka, Jira, Mantis, Bugzilla, etc. are well-known ones. Customized workflows are defined on this tool by providing different access rights to the design team to log, view, assign, resolve, and close the issues on them. Some design houses also use these to evaluate the quality of the designer and designer/ verifier. Typical example workflow for bug tracking is shown in Fig. 7.21.

7.17   Formal Verification

Conceptually, formal verification process is checking the response of the SoC design for all possible values of inputs with 100% coverage. This is highly impossible, to imagine the possible combinations of inputs, capture the response, and analyze them all. This is because of human limitations, computational resource limitations, and the time it takes to exhaustively verify the SoC design of complexity seen today. Hence, this is not generally practised in SoC design methodology. But the formal verification concept is used for checking the transformations the design undergoes during the design cycle to complement the verification of SoC

Fig. 7.21 Bug tracking
workflow

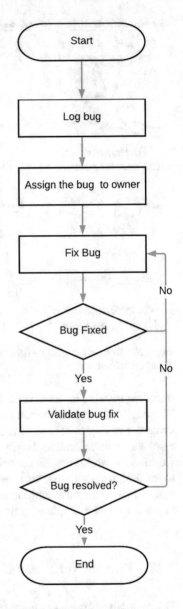

design by simulations. This is called equivalence checking. When the design undergoes transformations from RTL to netlist, equivalence checks are performed to
compare the netlist and the virtually synthesized netlist of RTL and are compared to
verify the equivalence. The RTL design is referred to as golden reference design,
against which the netlist is compared. During the design processes like synthesis,
place, and route stages (physical design flow), the netlist is written out and compared against the golden reference RTL design to check if the same design intent is
preserved by transformations. Well-known equivalence checkers are conformal
LEC, formality, sequential logic equivalence check (SLEC), and ESP.

Fig. 7.22   FPGA development platforms

7.18   FPGA Validation

To get first-time success of the SoC design is necessary to gain good confidence in the design that it works when fabricated, which is possible if you have a way to test it in the design form, which is closer to the hardware. FPGA platforms provide that setup for validation. Though these platforms are evolving to fit most complex systems, not every SoC can be directly ported on these devices due to the limited port list, memory onto FPGAs, or obvious speed limitations. So, FPGA-based validation is adopted in the functional verification methodology for functionally critical blocks or performance critical blocks. Major FPGA players, Xilinx- and Altera-based development boards are available for validation today. Another important advantage of having the FPGA validation phase in the design/development stage is that if most of the critical hardware can be ported on to the FPGA and with associated components on board, the development system can be used for early development of the software that can work on the final system on chip. A few of the FPGA-based development boards are collated in Fig. 7.22.

7.19   Validation on Development Boards

Further to gain more confidence in the SoC design, one can develop their own development platforms using all the discrete chip versions of the IP cores being used in the SoC and the FPGA for the customized blocks and validate the almost complete SoC in the design stage. Like FPGA platforms, these also serve as platforms for the early development of software that can be finally integrated on the SoC.

Chapter 8
SoC Physical Design

8.1 Re-convergent Model of VLSI SoC Design

VLSI SoC design flow involves stages where the design is converted to different forms until the time it is taped out for fabrication. These design conversions are from the design requirements or specification in document format to register transfer level (RTL) as behavioural model to design netlist in structural form (gate-level) to design layout as physical structures. The SoC design flow is seen as the re-convergent model with multiple transformations. The transformations of a SoC during the design process are shown in Fig. 8.1.

The final design database will be taped out (design file transferred to the fabrication house for further processing) as layout design file in GDS II file format. The design layout database is used to generate mask input data generation. An advanced software converts the complex design layout data to machine-readable commands for e-beam or other laser equipment used for pattern generations used in making masks. Mask databases include mask patterns corresponding to design structures, structures for pattern inspection, and metrology and data patterns directly writable onto wafers during SoC fabrication. It includes e-beam and photomask correction data to compensate for the differences between design mask patterns generated and actual patterns transferred on the masks. Mask data preparation is the first step in device fabrication. Mask data is used for mask making or for reticles. The difference between the mask and reticle is that a photomask is a pattern-transferring device onto complete wafer, but the reticle transfers patterns onto a small part of the wafer which must be stepped and repeated to cover the entire wafer. Photomasks and reticles are used during different stages of fabrication processes for selective processing on wafers. CMOS device fabrication processes include chemical vapor deposition (CVD), ion implantation, etching, and physical vapor deposition (PVD). A brief note on mask making is given in the last section of this chapter. The SoC design requirement is captured as a specification document called a chip architecture document,

© The Author(s), under exclusive license to Springer Nature Switzerland AG 2022
V. S. Chakravarthi, *A Practical Approach to VLSI System on Chip (SoC) Design*,
https://doi.org/10.1007/978-3-031-18363-8_8

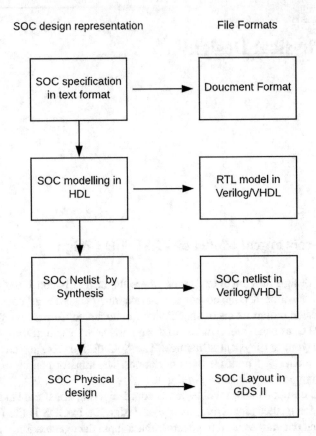

Fig. 8.1 SoC design representations

which is modeled using hardware description languages (HDL) such as Verilog/
VHDL or SystemVerilog, and then synthesized to gate-level netlist during logic synthesis which is followed by a process of physical synthesis and then routed with
interconnecting all design elements and saved as design layout in GDSII format. The
transformations during design can be explained using re-convergent models. At
every stage of the design transformation, the netlist is extracted and compared with
the original input netlist model for equivalence checking, i.e., to check if the design
intent is retained. Figure 8.2 shows SoC design re-convergence.

8.2   File Formats

During the various stages of design transformations, the design database is stored in
different file formats. Table 8.1 lists various file formats and their relevance.

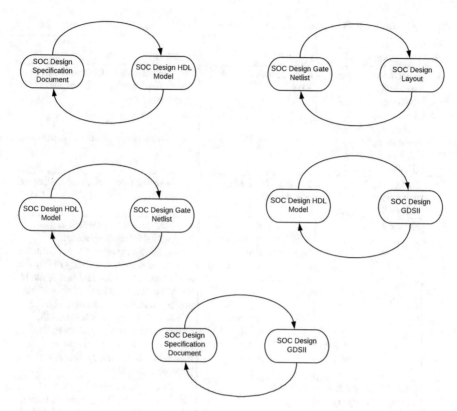

Fig. 8.2 Re-convergent model of SoC design

8.3 SoC Physical Design

SoC physical design is the process of converting the SoC design netlist to design layout and generating a design database in (graphic data system) GDS II format. Physical design is also known as the place and route (PNR) flow of the design. Physical design is an EDA tool-dependent and computationally intensive process typically carried out on high-performance, high-speed workstations. The physical design automation tools are required in design planning, with design exploration at the physical level, placement and optimization, clock tree synthesis, routing, manufacturing compliance, and fabrication signoff closure challenges. Tools are required to optimally place and interconnect many millions of transistors along with power and clock feeders in overnight runs. There are very sophisticated EDA tools that are used for PNR flow.

Table 8.1 Different file formats encountered in SoC design

Sl no.	Design stage	Format	Description
1	Requirement capture, marketing requirement document, architecture document, or high-level design (HLD) document	Document in docs, doc, XLS	Chip architecture is documented from market requirement, standard, and feature list
2	Design modeling using hardware description language	Verilog/VHDL files: .v, .vhd formats	The SoC functional behavior is modeled using HDL
3	Synthesis	Gate level file in Verilog/ VHDL file containing logic gates and interconnections; .vg, formats. lib files	The SoC design is converted to gate level netlist by the process called synthesis using synthesis tool. Synthesis tool can also write out liberty timing file in the form of .lib. Liberty timing file is the ASCII representation of timing and power parameters associated with the cell at various conditions. It contains timing models and data to compute input-output path delays, timing requirements (for timing checks), and interconnect delays
4	Static timing analysis and signal integrity checks	SPEF file	Standard parasitic exchange format (SPEF) file is the IEEE standard format for representing parasitic data in ASCII format on interconnect in the design. This is used by the static timing analysis tool to compute path delays and for interconnect data for signal integrity checks
5	Physical design synthesis	Frame files, physical-only library, new data model (NDM) libraries	Physical design tools accept design files in native formats represented as new data model (NDM) models for placement and routing. The frame files are extracted from layout design as physical only library without timing or logic information. They represent only the structure, area, and size information.
5	Static timing analysis/ dynamic timing analysis	SDF	Standard delay format (SDF) is the representation of timing delays

(continued)

Table 8.1 (continued)

Sl no.	Design stage	Format	Description
4	Floor plan and placement, global routing, clock tree synthesis	DEF, LEF files	Design exchange format file written as .def file by place and route tool contains die size, logical connectivity, and physical location in the die. Hence, it contains floor planning information of standard cells, modules, placement and routing blockages, placement constraints, and power boundaries Layer exchange format (LEF) provides technology information, such as metal layer, via layer information and via geometry rules. The LEF file contains all the physical information for the design DEF file is used in conjunction with LEF file to describe the physical layout of the VLSI design
5	Power routing	Layout file, LEF file, DEF file, lib file	
6	Detail routing	Layout file, LEF file, DEF file, lib file	
7	Tape out	Layout file in GDS II format	Industry standard database file format for data exchange for layout artwork. It is a binary file format representing planar geometric shapes, text labels, and other information about the layout in hierarchical form. GDSII file contains all the information related to SoC design. Once the design meets all the constraints for timing, SI, power analysis, and DRC and LVS, it means that the design is ready for tape out. This GDSII file is used by fabrication house for mask/reticle making

8.4   Physical Design Theory

A stick diagram is the basis for the physical design of the VLSI designs. Digital circuits are represented by a set of color-coded sticks and their relative positions. A stick diagram can also be in black and white colours with structural patterns in different layers of design elements represented as different shaded patterns.

8.5   Stick Diagrams

A sStick diagram is the method to capture circuit topology and process information with coloured layers in simple diagrams. They are the basis of layout representation of design elements of digital circuits symbolically. The stick diagrams do have notations and rules as shown in Fig. 8.3, and the coloured lines depict different layers which are also represented by different patterns of lines in black and white stick diagrams. Rules define the interconnection methods.

Few examples of stick diagrams for circuits are shown in Fig. 8.4.

As stated earlier, the stick diagrams have information of circuit device structures, their relative placements and interconnections of design elements and not exact coordinates. SoC physical design layout database has complete information of device structures, placement coordinates within the die layout, vias across the layers, and device interconnections. Mask data generated from the design layout is used for making masks or reticles for VLSI device fabrication. Mask or reticle facilitates exposing different parts of the die layout to different processes during fabrication. Mask is used to transfer the design data onto the reticle. The reticle is used to transfer the design structures on the silicon wafer and support step and repeat for production of a large number of dies on a wafer. Some of the processes during IC fabrication such as physical and chemical vapor depositions, diffusion, etching, ION implantation, and metallization make use of these masks or reticles.

SoC physical design process converts the SoC netlist to the SoC layout as shown in Fig. 8.5.

Complete SoC design conversion process is shown again in Fig. 8.6.

Detailed physical design flow is shown in Fig. 8.7.

During SoC physical design, material effects, process effects, and electrical effects of the fabrication processes are addressed to fabricate functional and reliable devices. Some of the issues such as electromigration effects of metals, coupling capacitance, and inductance effects, cross-talk effects, and IR effects are addressed in this design stage. Also, these design verification and analysis are done considering these effects. For example, timing analysis must be carried out and violation must be fixed considering the effects of interconnects, and electrical IR and antenna fixes are to be provided for seeing the issues of process effects. Physical design verification (PDV) is an important activity during physical design, which is dealt in detail in the next chapter. Over the years, this flow has been defined, refined, and time-tested as the physical design flow. The physical design tool or PNR tool consists of placer module, router module, CTS, and optimization and extractor modules which use the most advanced algorithms and also use analysis modules in conjunction to provide the layout of the desired quality of results (QOR).

Definitions of most commonly used terms in the physical SoC design layout are the following:

1. Track: The track is a virtual channel through which the P&R tool does signal routing in an SoC design. Tracks are defined for each metal layer in both preferred and non-preferred directions, which are used by the router module. The router routes the signal assuming the track to be at the center of the metal piece.

notations and rules as shown in Fig. 8.3, the colored lines depict different layers which are also represented by different patterns of lines in black and white stick diagrams. Rules define the interconnection methods.

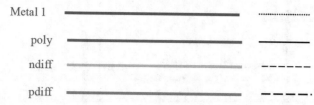

Rule 1. When two or more sticks of same colour touch or cross each other form a contact.

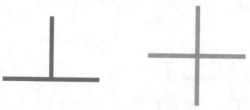

Rule 2. When two or more 'sticks' of different type cross or touch each other there is no electrical contact. If contact is to be represented, it has to be shown explicitly by a filled small circle.

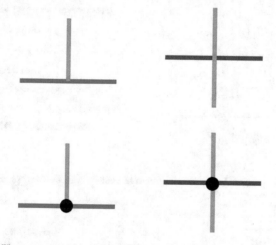

Rule 3. When two or more 'sticks' of different type cross or touch each other there is no electrical contact. If contact is to be represented, it has to be shown explicitly by a filled small circle.
Rule 4. In CMOS a demarcation line is drawn to avoid touching of p-diffusion with n-diffusion. All pMOS must lie on one side of the line and all nMOS will have to be on the other side.

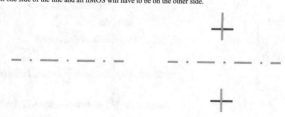

Fig. 8.3 Stick diagram rules

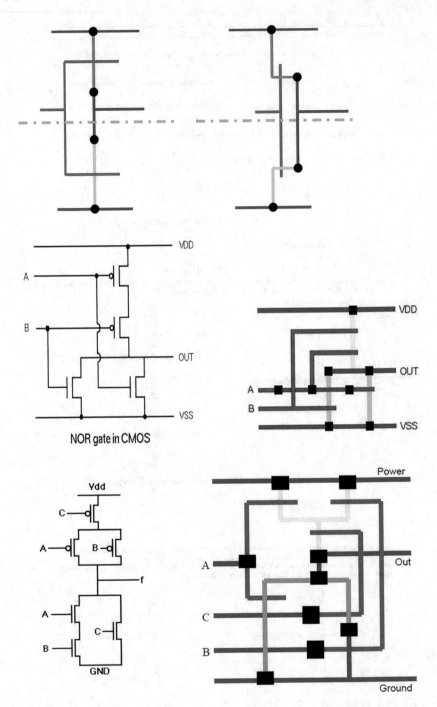

Fig. 8.4 Stick diagrams

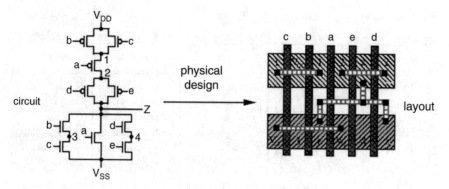

Fig. 8.5 Circuit representation and layout representation

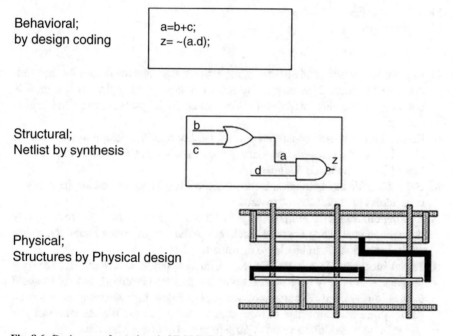

Fig. 8.6 Design transformations in VLSI SoC design flow

2. Row: This is the area defined for standard cell placement in the design. A row height is based on the height of the standard cells used in the design. There can be rows of various heights in the design based on the type of the standard cells used.

3. Guide: A module guide is the guided placement of a logical module structure in the design. The guide is a soft constraint. Some of the module guide logic can get placed outside the guide, and other logical module logic can be placed in the guide region.

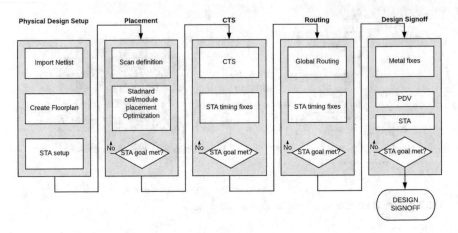

Fig. 8.7 Detailed physical design flow

4. Region: The region is a hard constraint in the design, and the design for the module is self-contained inside the physical boundary of the region. However, it is possible for outside modules to have some logic placed inside the region boundary.
5. Fence: This is a hard constraint specifying that only the design module can be placed inside the physical boundary of the fence. No outside module logic can be placed inside the fence boundary.
6. Halo: The halo or obstruction is the placement blockage defined for the standard cells across the boundary of macros.
7. Routing blockage: Routing blockage is the obstruction for the metal routing over the defined area. They are hard blockages in the design layout boundary, where no design element can be placed or routed.
8. Partial blockage: This is the porous obstruction guideline for the standard cell placement. It is very helpful in keeping a check on placement density to avoid congestion issues at later stages of design. For example, if the designer has put a partial placement blockage of 40% over an area, then the placement density is restricted to a maximum value of 60% in the layout floor plan area.
9. Buffer blockage/soft blockage: This is a type of placement obstruction in which only buffer cells can be placed while optimizing or legalizing. No other standard cell placement is allowed in the specified area during placement, but legalization and optimization of some cells can be placed in this region.

The physical design process of generating the design layout is very complex and can be studied under five headings: physical design setup, placement, CTS, routing, and the design signoff. In the design setup stage, the SoC design netlist is imported and floor plan is done after partitioning.

- The design setup during physical design involve following activities:

  - Appropriately partitioning the design suitable for their placement and inter-connections by routing.
  - The floorplanning.
  - Design environment setup for static timing analysis as per the design requirement with executable scripts as per the Static Timing Analyser (STA) tool.

- Placement stage involve the following activities:

  - Scan cell definition.
  - Placement of standard cells, module or block placements, macros, and the input-output (IO) cells.
  - Planning and routing of the power and ground distribution network for the design.
  - Reordering the scan cells.
  - Fixing design violations reported by static timing analysis.

- Clock tree synthesis (CTS) stage of physical design involve following steps:

  - Synthesising the suitable clock trees.
  - Fixing design timing violations resulted by clock tree insertions.

- Routing stage of physical design involve the following steps:

  - Routing globally the interconnects of design elements. This step is called the global routing. This creates routing channels for interconnects.
  - Fixing design violations for antenna effects; inserting spare cells, filler cells and corner cells.
  - Routing the actual interconnects in the design. This step is called the detail routing.
  - Optimisation of the design layout after routing the interconnects. This is called the post-layout optimization.
  - ECO flow for any lately detected design issues. The issues can be functional or timing violations. This step may use the inserted spare cells or doing alternative routing of interconnections.
  - Fixing timing violations resulted due to the routing, ECO fixing steps.

- The design signoff stage involve following steps:

  - Fixing routing issues in the metal interconnections.
  - Verification of SoC design for the original intent which could have disturbed due to the P & R design processes. This is called the physical design verification. This step consists of the following:

    - Finally fixing the STA issues.
    - Check the design for electric rules and resolve any violations. This is called the electric rule check (ERC) stage in designflow.
    - Check for process design rules and fix any of the violations in design. This is called the design rule check (DRC) stage .
    - Check for antenna and IR rules and fix all the violations if any.

8.6   Physical Design Setup and Floor Plan

Physical design (PD) starts by checking all the inputs to the PD. The inputs are design netlist, design timing, power, and physical constraints, and technology libraries. When the design inputs are released for physical design, the design database is analyzed with all the constraints specified. The SoC design will have different constituents like analog core and digital core memories in the form of macros or independent entities. Design is partitioned again if required, by considering placement requirements. The main considerations during design partition for physical design are design blocks using the same power supplies (same power domains) and proximity to the blocks interfaced and blocks with common design rules. Macros are placed with special care in terms of guard bands (blockages) as per their guidelines and accessibility to the interacting blocks. All analog blocks are placed together so that they are supplied with common power and ground lines, and associated design guidelines are followed in terms of isolation and load considerations. On-chip memory macros are placed centrally considering ease of access by multiple blocks accessing them. The external memory controllers such as the DDR controller are placed near the layout boundary near the input-output pads for easy connection to IOs.

Each of the partitions in a design is a netlist, that can be laid down by the designer using a PNR tool. Hard macro in design is retained as it is. Some of the main considerations for partitioning the design-netlist during floor planning are the following:

- The design netlist is decomposed into smaller netlists of sub-blocks without affecting the overall design functionality.
- Each of the design block gets connected to other blocks with minimum lengths of interconnections. This ensures that the interacting blocks are planned to be placed nearby.
- Maintain the minimum delay of circuit paths when the timing paths run across multiple blocks meeting the timing requirements of design elements. For example the input-output delays of the interfaces of interacting blocks must be kept as minimum as possible.
- Maintain the minimum number of signals interfacing the blocks and limit them to less than the set limit of maximum number of signals.
- Have subblocks of almost same area while partitioning the design. The estimated areas of the sub block must be within the limit specified by the process technology. A large number of subblocks of small size result in the easy physical design but increases the cost of the fabrication.

It is a trade-off between physical design complexity and the fabrication cost of the SoC design. Netlist is partitioned manually for time-critical part of netlists and others using the tool. The design tool uses iterative probabilistic algorithms for dividing the netlist into smaller blocks. Design partition is assessed for PPA with cost goals and improved upon incrementally again and again till it is partitioned to a level manageable by the physical design tools and designer. The partitioning algorithms running in the tool must be fast enough so that it is a small fraction of the complete physical design time. A few terms used in the design partitions are *terminal pitch* and *terminal count*. Terminal pitch is the minimum spacing between two successive terminals of the design block. This depends on the design rule as per the target technology. The terminal count is the ratio of the perimeter of the partitioned block to its terminal pitch. The number of signal nets which connect one partition to other partitions should not be more than the terminal count of the partition. During partitioning, critical components are to be placed in the same partition hierarchy. If that is not possible, a suitable physical constraint must be fed to the tool to place them closely. The constraints used for partitioning include area constraint and terminal constraint. The cost function for the partition algorithm is the number of interface signals that cross partition boundaries and the number times the signals cross partition boundaries. Figure 8.8 shows an example of a design partition.

The SoC physical design can be pad-limited or core-limited. If the number of pads is too many, it is pad-limited. On the other hand, if pads are too few and there is space between pads, it is core-limited. Chip planner module in the PD tool decomposes the SoC netlist into partitions which can be managed by the placer module of the physical design tools with proper constraints. This results in minimizing the interface signals across the partitions of the design netlist. Typically, auto planner require information about the area utilisation, average aspect ratio of subblocks and number of instances in the SoC design.

The area utilization is the ratio of the area which can be used to plan the design structures to die size estimate in percentage, the aspect ratio is the ratio of die height to die width, and a number of instances are a number of partitions in the SoC netlist. The partition tool will also restructure the design netlist to be able to structurally partition the netlist without breaking the functionality. After the design partition, planner module of the P & R tool can write out the netlist that is to be checked for equivalence with respect to the input synthesised netlist to confirm that the

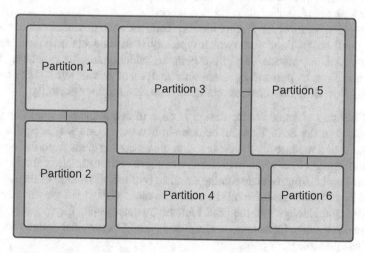

Fig. 8.8 Design partition example

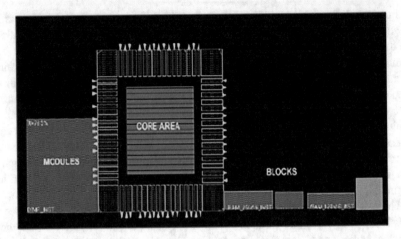

Fig. 8.9 Display screen during floor plan

functionality is not disturbed. Figure 8.9 shows the placement of sample SoC which consists of analog block, DDR controller, common on-chip memory, the digital core with processor peripheral subsystem core, etc.

8.7   Floor Planning

Floor planning of the SoC is an important phase of physical design where the location, size, and shape of the functional design blocks in soft (netlist phase) and hard macros are decided. If the design is analog, custom, or mixed mode, floor planning can also include row creation, I/O pad or pin placement, bump assignment (flip chip), bus planning, power planning, and more.

Floor planning involves placing blocks, modules, and submodules according to the prepared rough floor plan (which typically is in thoughts or paper). Present day EDA tools of provide VLSI floor planners which guide floor planning based on AI algorithms by performing space-time analysis of previous designs flows. The following flow describes the most common sequence for floor planning:

• SoC die size estimate for the design is done to determine the approximate PR boundary of the SoC. This can be done in two ways: One way is by listing all the cells and modules in the design with the estimated areas from the synthesis tools using unit area given in the library to arrive at the estimate of total cell area on the layout. Approximate routing estimate (typically, 30–35% of the logic cell area) is added to get the total die size. The other way to get the die size is by importing the design into the P&R tool and by determining the fitness boundary by repeated trials. Or use planner module in the EDA physical design tool to get the best possible floor plan.

- Placing the standard cells, partitioned blocks, macros, and IO cells of the design netlist.
- Initial floor plan is done by the P&R tool. This gives a good indication of how the blocks should be placed along with the orientations and grouped together in the die area within place and route (PR) boundary. This is repeated to get the right position of the design elements in the floor plan without overlaps and congestion.
- The design elements are placed for trial or virtual route is run to see if there is any overlap among them or design congestion. Optionally, the core area is resized at the block or module or die level to fit them. This is used as the guideline to do the final floor plan by the physical designer.
- The placer module of the tool is then used to place all the miscellaneous logic elements such as design wrappers, power, and ground cells.
- Sometimes, the floor plan of the critical design blocks of any design hierarchy and macros are generated separately. Iteratively, estimated die size, position and orientation of design blocks, alignment and placement density and optimum size is determined.
- The static timing analysis(STA) is carried out at this stage and every step from this stage as the design advance in physical design. The STA is carried out on the extracted netlist with extracted physical and timing constraints in every step from now on. Any violation is fixed in the design.
- Plan of distributing the power distribution to all cells is carried out in this stage.

8.8   SoC Power Plan

The floor planning of the SoC design also involves a power planning. To power all the design elements (standard cells, macros, and IO pads), it is necessary to plan the power grid. The power grid runs all around in the pre determined reserved paths called channels on the design layout such that the power supply to design element is tapped from the nearest point on the grid network. The power grid ensures the even distribution of the power supply to all the design elements. The power grid consists of the rings, stripes, and the rails. The *ring* carries power(VDD and GND) signals around the die. The *stripes* connect VDD and GND from the ring to the *rail* connect the VDD and GND signals to the standard cells as shown in Fig. 8.10.

If the SoC design requires multiple power supplies, many power grids are planned. The power specification is defined carefully as it is the lifeline of the SoC and is verified at every step to ensure every design element is well powered, before the design elements are interconnected.

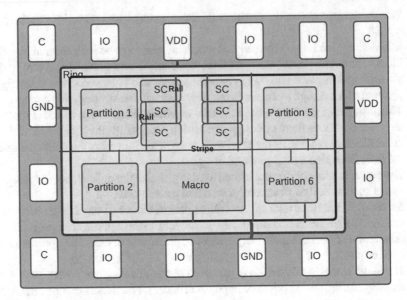

Fig. 8.10 Power grid plan

8.9   Two-Step Synthesis of SoC Design

The design netlist and constraints are written out after floor planning. The design netlist is used for logic equivalence check with the synthesis netlist as a reference netlist to ensure that the process of the floor plan has not disturbed the functionality. Any nonequivalences must be corrected so that the intended functionality is retained in the design.

In two-step syntheses, the design constraint with floor plan information is used for resynthesizing the design. This will help to meet the PPA goals of SoC design easily. The floor plan is repeated with the newly synthesized netlist.

8.10   Placement

The floor plan for placement is finalized if it seems to be optimum considering the feasibility of best interconnection of the design elements side. In this stage, the standard cells are placed in the rows created during the floor plan and the IO cells are are placed in the IO site or IO area of the design layout. To identify the correct orientation of larger blocks or macros, the *fly lines* are used. The fly lines show the connection of the macro or design component under consideration with the rest of the logic in the SoC design. This gives visual help to find and optimize the interconnects by placing them iteratively close to the blocks they are connected.

The steps involved in placement are the following:

Preplacement: In this stage, buffers and antenna diodes are added at the macro and block-level IO ports. Tap cells are added to avoid latch-up problems in the design. Spare cells are added to make the design failproof with metal corrections required in the further stage of the design process. Spare cells are used for last-stage changes in design as a part of the electronic change order (ECO) flow to fix timing or functional corrections.

Coarse placement: This step analyses the feasibility of routing to identifying congestions and probable timing issues in the design by performing the preliminary routing of the design cells. It is also used to identy the suitable locations and orientation for each of the design cell.

The legalization stage of placement of cells position them right considering the geometry and layout rules of the design. This may create violations of design timing paths, that are to be fixed. This is an iterative process where incremental optimization of the placement takes place until the design is entirely placed in the core area of the layout without the timing or design rule violations.

The high fanout net (HFN) synthesis step performs synthesis of high fanout nets (HFN) except for the *clock* and *reset* signals in the SoC designs. Control signals like chip-enable, read_enable, write_enable in processor-based subsystem designs are the examples of the HFNs. An HFN signal drives a large number of load cells. There is a limit on the maximum number of load cells, an output net can drive in terms of process technology. If any signal exceeds this number, it is called HFN. The tool in this step, automatically adds the buffers in the path of these signals while routing. The designer can declare an HFN as a don't touch signal in the constraint so that auto buffering is not done for such signals by the tool. However, it is necessary to add buffers manually for HFNs to avoid functional failures. The clock signal in SoC designs is an HFN. but, it is synthesised separately during clock tree synthesis as it is critical signal affecting design timing. For this to implement, clock signal has to be declared as idle for not treating clock signal as the HFN. If the clock is declared as idle signal, it will not be considered as an HFN, and buffers are not added to its interconnect paths.

In the placement optimization stage, all the placed cells are incrementally adjusted for their positions to avoid further violations. The congestion analysis is carried out for each of the design components. The defined blockages are used for the congestion analysis during placement optimization. Placement blockages are classified as *hard blockages*, and *soft blockages*. The hard blockage is an area of the design core area where placement of the standard cells is forbidden. Soft blockages are areas restricted to only a certain number of the standard cells or specific types of the standard cells depending on the process technology. Even areas around the places where the standard cells are placed can be blocked for further standard cell placement to avoid routing congestion.

Scan chain reordering: The design netlist from the synthesis will generally have connected scan flops. During the placement process, the scan chain connections order may get disturbed to the great extent. Sometimes, the connections can be very lengthy. It is therefore required to reorder the scan chains after the placement. The scan reordering step after the placement optimizes scan chain lengths in a SoC design to guarantee the routability. Scan reordering also helps to reduce congestion by reducing the interconnect lengths, thus reducing the number of repeater stages in its path.

Placement and optimization: The physical design (PD) tool helps the designer by features like the auto routing and resizing of the functional blocks, keeping relative placement intact. The PD tool is used to do the fitment trials on the design blocks by placing and adjusting the orientations without disturbing the interconnectivity, and even resizing the design blocks, to arrive at an optimized die size for the chip. Preliminary design analysis for congestion is carried out in this step. There are two types of the design placements supported by the P&R tools. They are *congestion-driven* and *timing-driven* placements. In congestion-driven placement, the routing congestion is relaxed during logic placement in the layout, at the cost of slightly higher interconnect length and overall silicon area. In timing-driven placement, the tool tries to achieve best possible timing for the design. There can be placement congestions, that need to be resolved. Major activities performed in placement stage are the following:

- Placement of special cells called spare cells (a set of extra logic cells of all types added to fix minor issues found during post-fabrication validation by metal tape out), endcap cells, de-cap cells, and JTAG cells close to IOs.
- Reordering of the connections of scan cells.
- The congestion-driven or timing-driven placement and optimization.
- The High fanout net (HFN) synthesis: HFN are signals like reset and chip enables which are required to drive large load or have high fanout. These signals are to be treated with extra buffers or cells of high drive strength to be able to drive the load correctly.
- Power distribution network (PDN) is generated by ensuring all the design elements are properly powered. The PDN typically consists of power ring, corner cells, power rails, and stripes to connect to the power pins of the design elements.

8.11   Physical Design Constraints

The size of the SoC design layout is initially calculated by importing design-nelist into the P&R tool. For preliminary estimation of the logic core size, standard cells and hard macros are considered alike. However, it is possible to determine how densely objects can be packed by weighing the standard cell density separately from the hard macro density: the standard cell density core size = (standard cell area/cell utilization) plus macro area plus halo. For fences and regions, effective utilization (EU = %) value is used. The EU value takes into account the actual cells and hard macros in the fence region, placement or routing blockages, partition cuts, and other floor plan constraints. It is a good practice to have right EU value before running placement. Placement of design is finalised when it is optimum in size. As a good

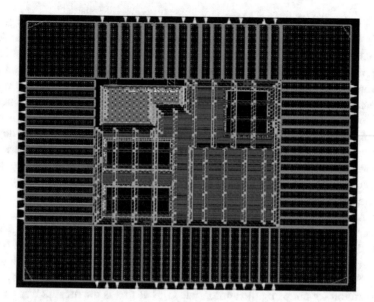

Fig. 8.11 Placed SoC design

practice, the hard macros are manually placed with appropriate orientation and alignment to get the optimum core size. The typical display screen after design placement is shown in Fig. 8.11. The modules are to be placed in the core area with desired orientation and location as in figure. STA is carried out and if there are no violations, design is considered ready for next stage which is clock tree synthesis.

8.12 Clock Tree Synthesis (CTS)

Clock tree synthesis (CTS) is an important design step in SoC physical design. 30 to 40% of the chip power consumption is because of the dynamic power consumption by clock circuitry and good clock architecture supported by clock gate, and clock tree implementation reduce power consumption and can yield good design performance. No clock signal generated is ideal and there will be uncertainties and variations in signal times. Clock uncertainties can occur due to many sources. It could be because of the clock generation logic, the device abnormalities in its path, power supply variations, interconnect effects, variation in operating conditions, load variations, and coupling effect due to adjacent signals. In spite of all these uncertainties, it is expected that with respect to clock, other signals such as data meets the setup and hold requirements of the sequential design elements for correct functioning. Hence, the goals of good balanced CTS, that meets rules of clock tree, design rule checks (DRC) by minimizing the clock uncertainty and meet the performance specification of the design. Most physical design tools are good at synthesizing the clock trees meeting the specified constraints, it is good to have manual intervention to check and fix any residual timing violations and design rule violations (DRV). It is an iterative process. Considering the criticality of clock signal in the design, it is essential to check design for timing violations and DRV violations. Additional checks are required on the derived clocks to identify the critical clock transitions, effects of parasitic capacitance in large fanout areas, congested or dense areas. The design is to be checked for correct drive strengths on high fanout nets and for clock balancing requirements. The CTS architecture is chosen, depending on the default rules set for the process technology. Clock buffer and inverters for clock tree are selected suitably considering, clock transitions, capacitance loads, and fanout values. While doing CTS, it is necessary to know clock structure and balancing requirements of the design by knowing the physical placement of the sequential elements. This helps to arrive at optimum clock tree for the design. Also it is necessary to know the design for the requirements of shielding the design areas, the need for fast clock transitions, areas requiring the cells with maximum capacitance, and the those driving the maximum load, so that balanced CTS, with appropriate buffer/delay cells are synthesised. Clock power consumption is also a consideration for choosing the appropriate CTS as this is the most power consuming network in the design. CTS use clock buffers and inverter cells with equal rise and fall times on input and output clock signals. If the clock network require boundary cells for a module or block, and then appropriate boundary cells with the correct buffers, and clock pins are added. These design details are rightly fed as constraint to the CTS tool. A boundary cell is a fixed buffer that is inserted immediately after the boundary clock pin to preserve the boundary conditions of the pin in the design. Boundary cell cannot be moved or sized. Also, no other cells are inserted between a clock pin and its boundary cell as shown in Fig. 8.12.

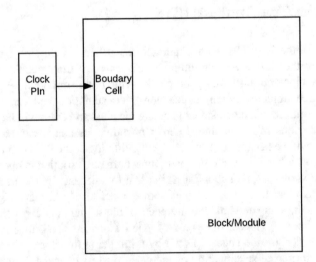

Fig. 8.12 Boundary cell insertion to preserve the boundary conditions

CTS run on the SoC design need clock tree design rule constraints which contains definitions for maximum transition, skew requirements, maximum capacitance, and maximum fanout. If the SoC design has multiple independent clocks, separate trees are to be built independently for each of the clock, in which case the CTS tool provides options to selectively block the tree on particular clock pin. This is possible by adding "Don't_touch subtree" like options in the constraint as shown in Fig. 8.13. This preserves a portion of the subtree untouched.

The synthesised clock tree in CTS stage is optimised to get balanced tree. The CTS module in the P & R tool does this by resizing the cells (changing buffer cells of optimal drive strength), relocating the buffer cells, gates, and shielding. Typical CTS on design is shown in Fig. 8.14.

Fig. 8.13 Don't touch subtree

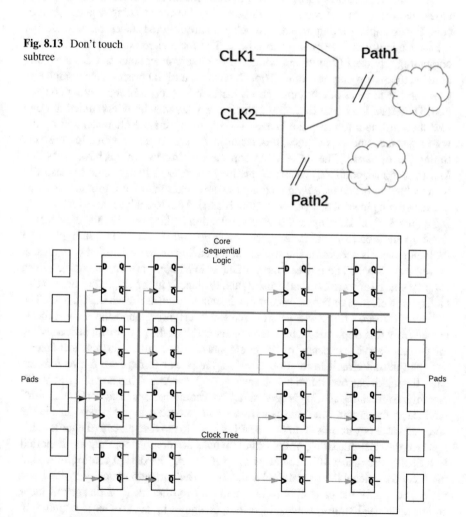

Fig. 8.14 Clock tree synthesis (CTS)

8.13   Routing

The next step after CTS is interconnecting the design elements by a process called the routing. This represents the metal interconnections of design elements in the SoC design layout. In the routing step, the input-output of design elements are connected as per the design netlist and is rewritten as a updated netlist. An advanced algorithm of the P&R tool is used to wire them one by one by metal as per the connections in the design-netlist. The routing algorithms are often based on "random walk" like algorithm where lines move from one grid to the other in random fashion but in a particular direction.

In a SoC design, the routing is done at different levels. All the complex design blocks requiring manual routing are done to start with the routing as a design process. For example, routing in analog block is a part of in the analog physical design is independently and manually carried out. The manual routing is the process of computing the sizes (length and width) of the interconnections and are manually drawing them using layout editor. Though it seems a trivial process, the complexity grows as the number of interconnections increase that result in the routing congestion. Therefore, the metal routing is done in many layers with wires routed through *vias* through them to cross the layers. The interconnects are characterized by the wire resistance and capacitance, that result in the signal delays, affecting the SoC timing performance. The metal interconnects running in parallel have cross-talk (electromagnetic coupling) when they are long on the SoC die. This issue is resolved by a technique called *shielding* techniques or by maintaining a minimum distance between them physical design verification is carried out to analyse the effect of routing on the design. The physical design verification is discussed in the next chapter. clock. In addition to signal routing, the power ground (VDD GND) routing is done through channels across the die so that all the design elements are fed with closest power ground pair. The power supply VDD and GND are primary inputs fed from external source and are internally distributed to all cells in the chip. They are distributed on power rings or power grids if the design is large, as shown in Fig. 8.15. The power and ground rings will encircle the SoC design core and the connection is tapped from this ring. The metal interconnect width of the power ring is decided by the current carrying capacity of the interconnect based on the process technology and the internal circuit in the SoC core; As the power feeders are drawn into the cells, the width narrows down just to carry enough currents. This is called line width tapering. Amount of tapering is determined by a combinations of factors and is summarized in the layout rules and is used for automatic routing of power. Inside the core, alternate power ground lines are laid as grids to tap the power to the logic cells.

Once the design routing of the critical blocks are completed, they are imported as a library file that will be routed at the SoC level. The design routing of digital blocks and top-level integrated SoC is finally carried out automatically by the P&R tools. Tools have the capability of automatically routing the SoC designs of large complexity. The list un-routed of input-output signals by auto router is written into a log file and are highlighted on the layout editor display for the designer to route

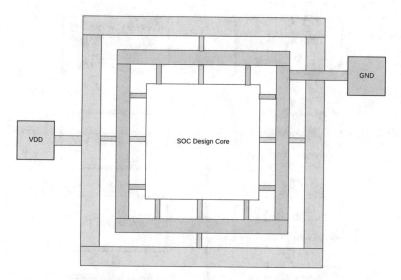

Fig. 8.15 Power ground rings

them manually. The designer for him to manually route them. In SoC design, the process of routing process is carried out as a two-step process called global routing and detail routing. In the global routing, the design elements which are easy to route are all connected, and in the detail routing, the tool performs auto routing of all the remaining signals (applying incremental routing steps iteratively) using alternative paths and tool times. SoC design layout is complete when all the signals, power, and clock are successfully routed. It is now ready for final test for manufacturability and tape out, provided it passes the design verification. After the completion of every stage of SoC physical design, viz., floor plan, placement, clock tree synthesis, and global and detail routing, the SoC design netlist and the updated constraint file are written for *logic equivalence check* with original reference design netlist file. The logic equivalence check (LEC) and the STA must pass to advance the SoC design at every stage of the physical design.

8.14   ECO Implementation

The design changes for fixing the functional and the timing issues in SoC design are inevitable even in last stages of design as the design verification continues till the design is taped out. Incorporating these changes in RTL is not practical when the design is in the advanced stage of physical design. The critical design changes are implemented during physical design as electronic change orders (ECOs). ECO files are small handwritten netlist level corrections or synthesized netlist used to fix timing issues or logic corrections. These are manually written netlist file or logic corrections in the design netlist file, that contains a specific set of standard cells and

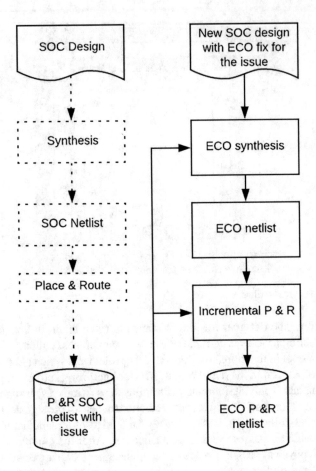

Fig. 8.16 ECO implementation flow

their interconnections. ECOs are acceptable in the SoC design as they generally do not change the performance of the SoC design except that they correct the issues found. The ECO file is imported into the physical design tool and the design is routed as incremental place and route on the SoC design. The ECO design flow in SoC design is shown in Fig. 8.16.

8.15   Advanced Physical Design of SOCs

Since today's SoCs demand extreme low power and high performance, it is necessary to control the otherwise automatic tool-dependent process of the physical design very closely with manual intervention. That means tool runs can be tweaked for the best performance of the SoC. The following sections describe some of the advanced physical design techniques adopted in SoC designs.

8.15.1 *For Low-Power Consumption*

Low power SoC design is the need of the hour. Physical design processes can be refined or tweaked to realise very low power SoC designs. PD for low power uses a combination of circuit design techniques, special cells in the technology library and the P & R tool configurations. *Power domain* or *voltage island* is a floor plan object, that has its own .lib and .lef files associated with it. By keeping the power domain as the floor plan object, it becomes easy to implement special design elements like voltage *level shifters* or *power gate* for realizing low-power designs using the physical constraints. The floor plans are implemented as multi-supply single voltage providing different levels of isolation or multi-supply multi-voltage domains. Reduction of power consumption is achieved either by shutting down a power domain or operating it at a reduced voltage (voltage scaling). Power domain shutdown is a technique in which an entire power domain is shut down during a specific mode of operation. This results in the savings of both *leakage power* and *dynamic power* in the designs. This is because the transistors in the functional blocks are isolated from the supply and ground lines and are switched off when not active. You must use isolation cells when shutting down power domains in order to drive the interface signals to predetermined known states. In many cases, a design in the shutdown mode operates at a single voltage throughout the design (an MSSV design); however, the portion of the design that is shut off must be in a different power domain. This is necessary because this portion must be isolated from the rest of the system so that it can be shut off independently from the rest of the core logic. In the power domain scaling (also known as voltage scaling), one or more domains operate at a voltage lower than that of the other core logic. The power domain scaling provides dynamic power savings, and may provide the leakage power savings, depending on the threshold characteristics of the library for the scaled domain. These power gating and voltage scaling techniques can be used separately or together in a design to achieve low power. These techniques require special power switch cell, on-chip power regulator cells, and level shifter cells.

8.15.2 *For Advanced Technology*

With the advent of the advanced technologies alternative to CMOS technology, the physical design tools also offer a wide range of flexibilities considering the fabrication processes involved in those technologies. Support for standard FinFET technology is explained in this section. FinFET device is the 3D structure compared to planar MOSFET transistor as shown in Fig. 8.17.

Compared to the MOSFET, in the FinFET devices, gate wraps around the diffusion FIN structure to gain more control on the channel current. This also promises higher performance in terms of the speed at the same power level as the planar MOSFET technology. Hence, designer can target higher speed for the same power

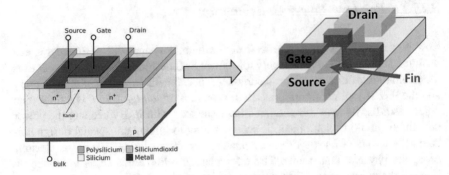

Fig. 8.17 Planar MOSFET and FinFET structures

level or same speed at low power as that of the MOSFET designs. This requires all the placable structures to be aligned to the FinFET grids to manufacture these devices. So, the physical design tools support the FinFET grids with the fin to fin pitch support and checks on the snapping to these grids and alignment of the placement of objects with them. The tools have option to load the FinFET technology grids if the target technology is to be supported.

8.15.3 High Performance

Some of the design blocks require manual intervention in the physical design to achieve high performance in terms of the power, area, and the timing for critical data paths in the SoC design. This is done using separate *physical design constraint* (PDP) constraint during placement stage. In this case, the auto run flow is interrupted and manual intervention is permitted to implement the design as per the tighter data-path constraint for the specified performance manually by the designer. The placement congestion issues, alignment issues, orientation, and positioning are decided manually knowing the performance impact. The data-path design elements are read separately by the tool using a executable script and proper naming convention to identify them. The cells in preferred data path placement (PDP) are placed in separate area on the chip layout which is not considered for automatic placement. The PDP cells are used separately as a placable object with the permission to do the manual adjustments during physical design. The main advantage of the PDP placement is that it ensures uniform routing for the PDP. The PDP flow is shown in Fig. 8.18. Post routing the physical design verification procedure till the design tape out remains the same as traditional approach.

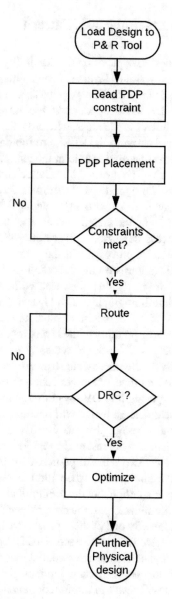

Fig. 8.18 SDP physical design flow

8.16   Photolithography and Mask Pattern

The main concept of SoC design depends on the possibility of creating the patterned material on a semiconductor wafer as layers for controlling the current flow. This is possible by the process called a photolithography. This enables transferring layout patterns generated by the EDA tools corresponding to the desired functionality onto the physical metallic structure onto the glass, which results in a mask or reticle. The minimum feature size talked about in the VLSI terminology ultimately depends on the resolution of the patterns that are feasible in the photolithography process. The design output in GDS II format is converted into Caltech Intermediate Format (CIF) which is used to create masks or reticles. The dimensions of the patterns on the mask or reticle will be many times larger than the actual desired patterns on the chip dimensions. This enables getting the finer dimensions on the wafer when processed. The photolithography process depends on the philosophy of creating transparent and opaque regions for selective processing of the planar regions on silicon wafers. The chrome-based metallic patterns on the mask reflect the light source making it opaque, while in the rest of the regions, the mask will be transparent to the light source, hence the name photolithography. Each layer in the SoC design layout will be transferred into a mask which is patterned separately. Hence, for a single chip, there may be 8–12 masks corresponding to the layers required as per the fabrication process. This is an extremely costly process, typically costing in the range of 500 K to 1000 K USD. This is due to the microscale structure required to be fabricated on the chip. The Chip fabrication process involves coating the semiconductor wafer by photoresist material and exposing it to UV rays through the mask so that the photo-resist will undergo chemical change which will become soluble in developer solution. This is similar to the photography process. By this process, the patterned regions are selectively etched and the rest of the regions are hardened, forming hard patterns on the chip. There are two types of photoresists: positive and negative photoresists. When in positive photoresist, when illuminated (by UV rays), regions become soluble in developer solution, but the unilluminated regions remain hard. In negative photoresist, the illuminated patterned regions are hardened while the unilluminated regions are soluble. By one of these processes, the hard patterned layer is formed on the chip and is selectively processed. This is repeated for as many layers as required in the design layout. It is hence essential that the patterns in the layout, during physical design, follow geometrical guidelines given for the fabrication process. Violating these rules will result in nonfunctional chips. The layout tools provide the ability to translate these patterns back into schematic again. This is required for the layout vs schematic (LVS) check to ensure accurate representation of the desired circuit. The tools also extract the circuit schematic from the layout drawings which include every electronic element and wiring detail as well as the parasitic resistance and capacitance of every line. This extracted parasitic RC file (wire resistance R and wire capacitance C file) is used in the verification of the electrical behaviour of the system on chip. On-screen layout structures from the EDA tool, mask pattern photolithography process, and patterned metal region on the wafer as an example of selective processing in IC fabrication are shown in Figs. 8.19 and 8.20.

For more information on detail fabrication processes, it is suggested to refer to CMOS VLSI design books.

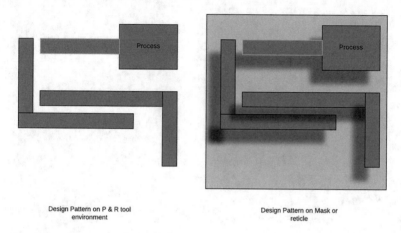

Design Pattern on P & R tool
environment

Design Pattern on Mask or
reticle

Fig. 8.19 Design pattern transfer onto mask

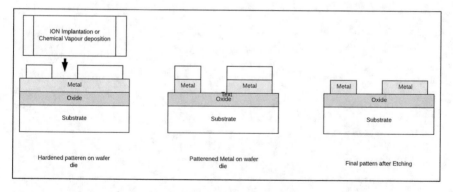

Fig. 8.20 Selective processing in CMOS fabrication process

Chapter 9
SoC Physical Design Verification

9.1 SoC Design Verification by Formal Verification

VLSI SoC design flow involves the transformations of SoC design from one file format to another during logical and physical synthesis. This is very well represented by the re-convergent model discussed in the last chapter. Debugging the SoC design in netlist format is feasible for smaller complexity but for bigger designs, it is extremely difficult to trace the root cause of the issues. Also, it is very difficult, very time consuming, and practically almost impossible to simulate the gate-level design for all scenarios. However, it is essential to check the design intent is retained in all these design transformations. This objective is achieved by alternative design verification methods such as static timing analysis and formal verification methods. We dealt with in static timing analysis (STA) in earlier chapters. Formal verification methods are dealt in this chapter as a part of physical design verification. Two types of formal verification methods are model checking and equivalence checking.

9.2 Model Checking

System modeling is a process of identifying the system properties and representing it as a set of mathematical equations or simulation reference models. SoC design is then verified by comparing it with the system reference model. This is done in simulation at the RTL level. When the design is in gate-level netlist format, it becomes very difficult to verify it by simulations. An alternate way is to derive a common database from the model and design and compare them for equivalence. For example, consider a coffee and tea vending machine and different machine states as shown in Fig. 9.1. The vending machine disperses coffee if a *coffee* button is pressed

© The Author(s), under exclusive license to Springer Nature Switzerland AG 2022
V. S. Chakravarthi, *A Practical Approach to VLSI System on Chip (SoC) Design*,
https://doi.org/10.1007/978-3-031-18363-8_9

and *Rs. 15 is inserted*, and tea if a *tea* button is pressed and *Rs.10 inserted* in the coin slot. The function of the vending machine is modeled by logical equations by formal methods. Different states of the machines are modeled.

Both the design and model are represented in Kripke structure and the properties are represented in the temporal structure time-dependent (state machine) which are input to the model checker and are compared for equivalence. A Kripke structure is a graphical method to represent behaviour of the system, named after its inventor Saul Kripke. It is a graph whose nodes represent the reachable states of the system and edges represent conditions for transitions of the system state. The model is treated as a golden reference against which the system design is verified. Any non-equivalence is considered as a design issue and must be fixed. The concept is shown in Fig. 9.2.

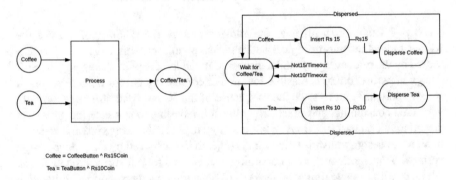

Fig. 9.1 Coffee/tea vending machine, state diagram, and formulae representing formal properties

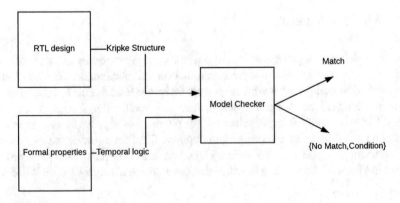

Fig. 9.2 Model checking

9.3   Logic Equivalence Check (LEC)

Logic equivalence check involves using the synthesized design as the golden reference against which transformed design extracted from EDA P&R tools is compared for logical equivalence. The concept is explained in Fig. 9.3.

The method involves converting the reference design and the design to be checked into netlist format using techniques like virtual synthesis, or by mapping physical structures to device logic cells and interconnections and finding the equivalence between two netlists. It involves comparing the netlist file extracted from the physical design and the original design netlist generated by the synthesis process to establish the correspondence and equivalence. The step-by-step process of LEC is shown in Fig. 9.4. LEC is run at different stages. RTL versus gate level netlist LEC is run after synthesis stage netlist vs reference netlist LEC is run after placement and after design routing.

LEC tools have a very user-friendly graphical user interface (GUI) for design debugging to understand the cause of nonequivalences. The tool cross-references the nonequivalence points to the schematic and source code so that designer can trace the logic path and fix the design issue. The tools can also be guided by a set of commands mapping the compare points manually by specifying the same naming conventions. LEC is run to compare RTL design and netlist or netlist versus reference netlist. Design is verified for logical equivalence whenever the design is changed during the design flow.

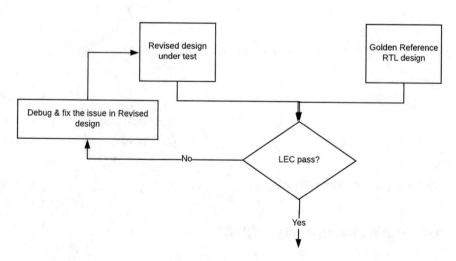

Fig. 9.3 Concept of logic equivalence check

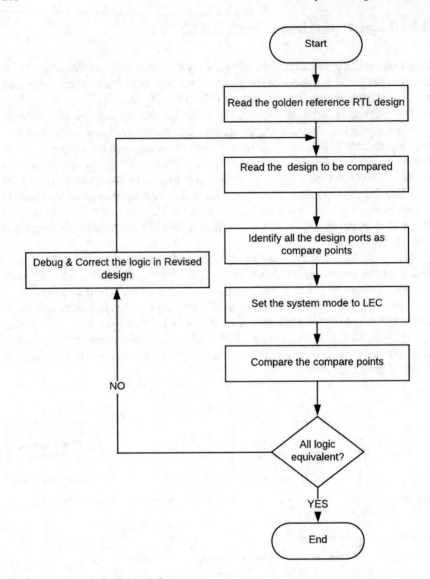

Fig. 9.4 Logic equivalence check flow

9.4 Static Timing Analysis (STA)

Static timing analysis (STA) is covered in Chap. 5. Timing analysis is an important physical design verification technique. It is run whenever the design is changed, either at placement or routing or after the electronic change order (ECO) stage. It is also run as a part of the design signoff procedure after the final layout with metal fills and antenna fixes are done. The parasitics, the interconnect resistor, and capacitance (R and C) values, are extracted from the actual physical design. These are fed into the STA tool along with the design files and timing library and the design timing reports are generated. These reports are generated for all the modes the SoC design operates for all the design corners. The STA tool selects appropriate values from the timing library file (whether it is best BC, typical TC, or worst-case WC) and reports the design timing paths for BC, WC, and TC conditions. Each of the reports is to be analyzed for any negative slacks and appropriate fixes are to be done in the design. Many PD tools have the capability of using these reports from integrated STA and optimizing the design for the set speed performance in the design timing constraint. Only a set of residual timing paths are to be fixed by the designer which is beyond the scope of PD tools. If advanced timing libraries corresponding to on-chip variations (OCV) are provided, PD tool optimizes the design considering OCV during the process.

Apart from the abovementioned timing analysis, it is necessary to analyze the design for skew, pulse width, duty cycle, and latency. The design netlist is read by the STA tool and the violation report is written out. If there are violations, they are fixed by adjusting the path delays in the gate-level netlist and running STA again. It is an iterative process. The design extractor modules in PD flow extract parasitics delays in SPEF format and design netlists. Once all violations are fixed, the SDF file is written out from the tool to use in the gate-level simulation. The gate-level simulation can be run to complement to each other running early by manipulating the SDF file written out of STA. The flow is shown in Fig. 9.5.

9.5 ECO Checks

SoC design changes with ECO implementation have to be verified for logic equivalence and static timing by utilising of LEC and STA techniques as explained previously.

9.6 Electromigration (EM)

Interconnection inside the chip generally uses aluminum and of late copper. Aluminum and its alloy interconnect lines exhibit a phenomenon called electromigration. These are typically found in the supply and ground rails which always carry unidirectional current. Electromigration occurs after years of usage of the SoC. When constant current flows through the power and ground interconnect for a long time, ions get knocked out by electrons from one place to another creating piles of ions at one side called hillocks and consequently void at the other end. This results in open/short faults on the interconnects. This can lead to reliability issues in SoC. Electromigration rules are added as electrical rules which have to be adhered to avoid such failures. There are three types of electromigration rules: DC,

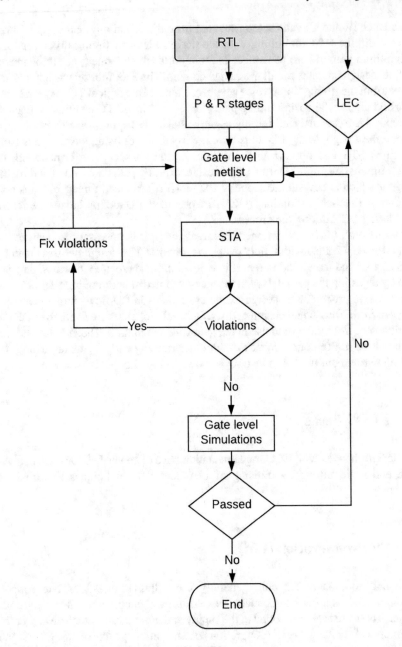

Fig. 9.5 STA and gate level simulation during PDV

time-varying unidirectional flow, and bidirectional AC. These are considered while designing power grids. The verification of these design rules is to be passed before the chip tape-out. Electromigration is not seen much in copper interconnects and hence gaining importance in today's SoCs. Copper as interconnect does not exhibit the problem of electromigration.

9.7 Simultaneous Switching Noise (SSN)

Simultaneous switching noise (SSN) is another problem seen in high-frequency SoC designs, if not taken care of. It occurs when a large number of logic gates change logic states at the same time that can lead to system failures. When many logics switch simultaneously, the voltage fed to the circuit around them becomes a time-dependent function of the current. The fluctuating current, due to parasitic line inductance L (though small in value which can otherwise be ignored) existing on any conducting interconnect, increases voltage drop reducing the effective voltage at the circuit:

$$V_{\text{eff}} = V_{DD} - iR - L\frac{d_i}{dt}$$

Please note that this drop exists on the GND line as well as the power line and can double the effect. This dynamic change in V_{dd} has to be taken care of by considering the data flow in the design and carefully following the layout design rules for power grid design. Separating the pad ring power supply and logic core power supply is one of the ways for avoiding parasitic effects on performance and reliability. Also tapping power supplies from all sides of the die and evenly distributing low-frequency and high-frequency input-outputs are generally done to avoid interconnect effects on SoC performance. These rules are checked in the electrical rule checkers (ERC).

9.8 Electrostatic Discharge (ESD) Protection

Electrostatic discharge (ESD) is a critical factor in modern CMOS design. The ESD destroys the thin oxide of the transistor layer, thus inducing device failures due to input transistor failure in pads. This is very common in SoC designs if they are not addressed during design stage. However, the input pad structures often come with ESD protection circuit, shown in Fig. 9.6, which is simple reverse-connected diodes between input line and power supply structure connected to sink large ESD energy by Zener effect. Protection circuits must be added to pads to take care against ESD as shown in Fig. 9.6.

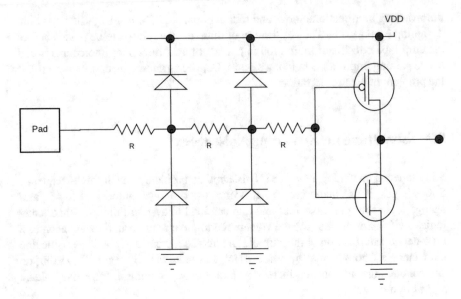

Fig. 9.6 ESD protection at input pads

9.9   IR and Cross Talk Analysis

Due to the high operating frequency of SoCs at multi-gigahertz, it is very essential to perform signal integrity (SI) and power integrity (PI) checks like IR drop, cross-talk effects, and noise to ensure first-time success of SoC designs. Noise effect on SoC can be due to the following reasons:

- Technology scaling resulting into high transistor density.
- Power supply voltage reduction less than 1 V.
- Increased switching and power density.
- Power supply noise due to resistance on power nets, spatial variations on power grids, and temporal variations of power supply voltage.
- Cross talk due to one signal interfering with another signal, capacitive cross talk between RC lines floating and/or drive nets on a chip floating, and signal coupling between nets due to LC transmission line effect.
- Inter-symbol interference.
- Thermal and shot noise.
- Parameter variation.

Static and dynamic IR analysis has to be done to check if the hotspots are within set limits so that they do not affect the reliability and performance issues for SoC. If not addressed properly, all of the above can render themselves as noise source leading to "hard to find" intermittent errors at current switching frequencies. So to curtail the effects of the same, good practices are translated into layout guidelines which are expected to be followed during physical design. One of the layout guidelines is to avoid floating nets which will result in capacitive cross-talk, picking up signals from neighboring nets. Layout guidelines will be stringent for sensitive circuits like low swing on-chip buses, dynamic memories, and low swing pre-charge circuitry near supply lines. The inductive effect functioning of input-output circuitry of mixed-signal and analog circuitry will not be pronounced in digital circuits. Congestion analysis is to be carried out and cell congestion has to be relaxed by suitably replacing them and distributing them accordingly. Also, the cross-talk effect is restricted by adding level-restoring circuits called keeper cells in dynamic switching circuits.

Few of the guidelines to be followed during design layout to avoid cross talk are the following:

- Avoid floating nets.
- Add keeper cells for sensitive circuits like pre-charge circuits.
- Separate or spread apart the sensitive net routes from fast switching nets.
- Do not have long interconnects on the same layer and parallel interconnect nets are laid with sufficient gaps.
- If required, shield the fast-toggling nets to protect the power (VDD) and ground (GND) nets to avoid IR drops. Dynamic IR analysis may show up hotspots due to cluttering of clock buffers in some spots showing up high switching activity. It must be taken care of by evenly distributing the clock buffers across the die.

The PD verification tools analyze the design and report violations if any for hotspot regions. These are to be fixed by the designers. Some of the design fixes for these issues are generally adhering to the design rules, spreading the logic apart, and increasing the width of metal interconnects by setting non-default rule (NDR) constraints on them. The final reports from this verification are used as signoff tools for accepting the design for fabrication. A lot of literature is available in VLSI books on interconnect effects in routing like RC modeling and parasitic parametric effects on electrical performance. Interested readers can go through them for extra information. Also, the tools explain in their user manuals how to run this analysis.

9.10 Layout Verse Schematic (LVS)

LVS is one of the traditional signoff techniques for taping out the design. Physical layout is checked for retaining the intent of the real design by extracting the netlist from the design layout. Design netlist is extracted from the structural content in terms of interconnections from metal layers and base layer structures and by checking for opens, shorts, and any overlapping base cells. Design extracted netlist is saved in SPICE format. Design extraction is very critical in this flow. The design extraction tool extracts the polygon structures to determine components like transistors, diodes, capacitors, and resistors and their connectivity information by identifying the layers of construction. Device layers, terminals of the devices, size of devices, nets, vias, and pin locations are defined and will be assigned a unique identification. This is further extracted in the form of a netlist. Once done, the layout netlist is compared with the golden schematic netlist of the design for preserving the design intent using an LVS rule deck. In this stage, the number of instances, nets, and ports is compared. All the mismatches such as opens, shorts, and pin mismatches are written out in the LVS report. Check is done comparing the number of design elements in the layout extracted netlist and netlist in the same stage using the LVS rule deck. LVS compares circuit topology and device sizes. Comparison is also done among the number of devices in the layout and schematic netlists, types of devices, and the number of nets in the two netlists. Typical errors in LVS checks are number of devices not matching with the number of devices in extracted view, shorts of nets that are not seen in the golden netlist, component mismatch, value changes in the form of sizes, etc. The physical design tools provide LVS feature with cross-referencing across the layout views for violations and corresponding netlist viewer to debug the issues. Some of the LVS errors are open net, short net errors, parametric extraction errors, and device mismatches, and their corrections are shown in Fig. 9.7.

There are dedicated LVS check tools that are used for signoff for tape out. The difference between the inbuilt LVS module in the P&R tool and the dedicated LVS checking signoff tool is that the latter uses all expanded details of physical structures of the design, while the latter uses abstract layout structures. The set of commands is written as an LVS rule set which is fed to the LVS checking module along with design file, layer assignments, and physical database checks like SNAP and GRID checks. Schematic netlist provides complete cell-level information along with nets. LVS flow is shown in Fig. 9.8.

9.11   Gate Level Simulation

Once all the STA violations are fixed in the design, gate-level simulation with back-annotated extracted timing information in SPEF format SDF is executed for sample functional scenarios. For the revised design netlist layout, the parasitic extraction tool writes out the timing information in SPEF file format, that is included (back-annotated) to run simulations. Note that the extracted design netlist file, require a library file, to be included to run the simulations. This is done by replacing the design under test(DUT) in functional test bench with the extracted design netlist file from any stage of physical design written out by the physical design tool and reading the SPEF file by back-annotation. This is called dynamic timing verification. The gate-level simulation with back-annotated timing is a tedious process because all the timing parameters like setup and hold times of design elements have to be correct to pass the vector. This will require fine-tuning the timing of input stimulus considering the design input latency specifications. Hence, starting early the gate level simulations for SoC design soon after synthesis stage helps to resolve any test bench issues. This helps to keep the gate-level simulation setup ready for the final dynamic timing sign off simulations run after all the timing violations are fixed in the design. Figure 9.9 shows the gate-level simulation flow for the time-closed SoC design.

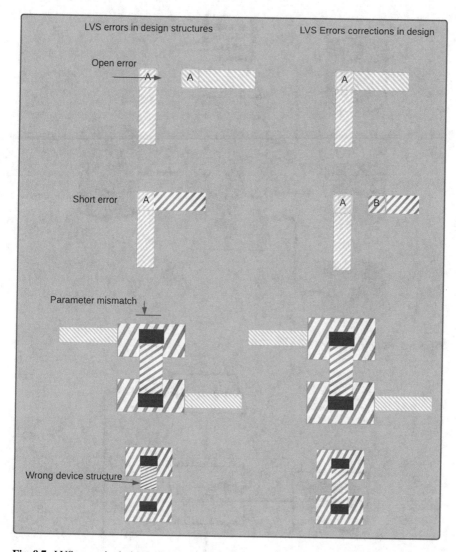

Fig. 9.7 LVS errors in design and their corrections

9.12 Electrical Rule Check (ERC)

Electrical rule check (ERC) is typically a static and dynamic IR analysis to detect IR drop bottlenecks, violations of electromigration (EM) rules, extensive checks for connectivity and reliability such as weak spots in the power grid, resistance bottlenecks (through short path tracing), missing vias, and current hotspots. The tool provides what-if scenario analysis on IR and EM by using region-based power assignment, so that designer can choose the right option. A typical IR map is shown in Fig. 9.10.

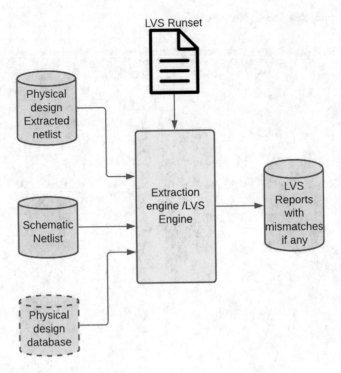

Fig. 9.8 LVS flow

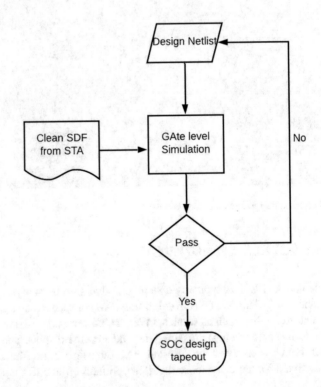

Fig. 9.9 Gate level simulation flow

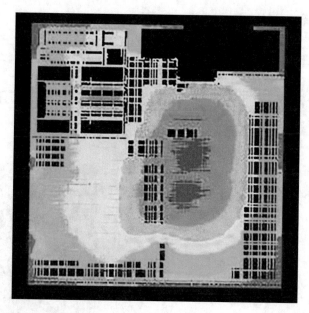

Fig. 9.10 IR map (Source: Celestry Design Technologies)

9.13   DRC Rule Check

Physical design is verified for design rule violations (DRV). This is done by the process called design rule check (DRC). It is a process of checking physical design layout data against the fabrication-specific rules supplied by the process vendor to ensure successful fabrication. Process design rules are related to the X-Y dimensions of layout structures and not the vertical dimensions of the layers. Example rules from foundries for a few technologies are shown in Table 9.1.

DRC is done by tools by generating computational geometry from the SoC design layout and checking the relation of overlap or distance between polygons of either the same or different layers. A screenshot of the DRC run is shown in Fig. 9.11.

Typical design rules for a particular technology node look like the one shown in Fig. 9.12.

9.14   Design Rule Violation (DRV) Checks

DRV is typically performed during physical design after CTS, routed, and as the final design completion. The typical steps involve the following:

- Perform RC extraction of the clock nets and compute accurate clock arrival time.
- Adjust the I/O timings.

  - After implementing the clock tree, the tool can update the input and output delays to reflect the actual clock arrival time.

Table 9.1 Sample DRC rules for different technologies

DRC rule	130 nm	90 nm	65 nm	45 nm
Width-based spacing	1–2	2–3	3–5	7
Min area rule	1 pitch	2 pitch	3 pitch	5 pitch
Cut number (via)	N/A	1–2	4–5	5–6
Dense EoL (OPC)	N/A	N/A	M1/M2	All layers
Min step (OPC)	N/A	1	5	5

Fig. 9.11 Screenshot of DRC

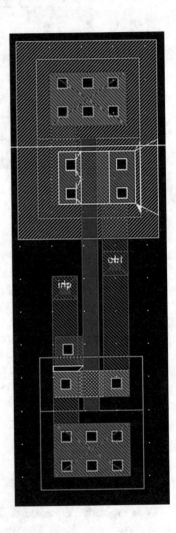

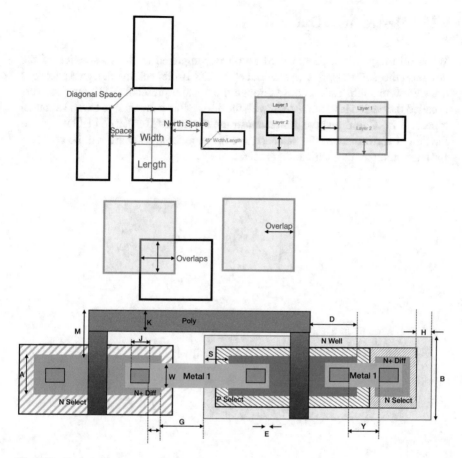

Fig. 9.12 Design rules

- Perform power optimization.

  - Use a large/max clock gating fanout during the insertion of the ICG cells.
  - Merge ICG cells that have the same enable signal.
  - Perform power-aware placement of integrated clock gate (ICG) and registers.

- Check and fix any congestion hotspots.
- Optimize the scan chain.
- Fix the placement of the clock tree buffers and inverters.
- Perform placement and timing optimization.
- Check for major hold time violation.

9.15   Design Tape Out

When all the physical design verification is completed to the satisfaction of the designer, the SoC design is written out as a GDS II file, and the design database is transferred to a fabrication house through file transfer protocol (FTP) process. This is called the design tape out. Along with the design file, it is required to tape out final reports of DRC runs and the design constraint file in SDC format so that DRM verification is performed on the database and if cleared, the database will be accepted for fabrication by the fabrication process.

Chapter 10
SoC Packaging

10.1 Introduction to VLSI SoC Packaging

VLSI SoCs have to be packaged such that they can interface with the rest of the world in a product to be used as a single unit or be interfaced with other circuits. They also have to be protected from mechanical stress, environment (humidity, pollution), and electrostatic discharge during handling. In addition, SoCs have to be exposed to be tested to ensure reliability with tests like the environmental test, burn-in tests, and other safety tests before they are ready for use. This is achieved by packaging it. Packaging provides a high-yield assembly for the next level of integration or interconnection on board for realizing the final product. Hence, the package must meet all device performance requirements such as electrical (inductance, capacitance, cross talk), thermal (power dissipation, junction temperature), quality, reliability, cost objectives, and testability at the package level. Hence, system on chip dies are assembled in the package. Major functions of packaging, therefore are the following:

1. Protecting the system on chip from environment and handling.
2. Provide path for Heat dissipation from chip to the ambience.
3. Provide reliable electrical connectivity to the neighboring systems through interface pins.
4. Packaging for handling further reliability tests on the system on chips.

Packaging and bonding wires on packages introduce inductive parasitics which can have an adverse effect on the SoC functioning. The current flows through input-output wires due to high signal transition activity which can cause voltage fluctuations like ringing, overshoots, and undershoot on supply rails affect SoC functionality. SoCs with more than 1000 pins can be seen today and designing a package for handling simultaneous signal variations to minimize the inductance effect is challenging. A few examples are shown in Fig. 10.1.

V. S. Chakravarthi, *A Practical Approach to VLSI System on Chip (SoC) Design*, https://doi.org/10.1007/978-3-031-18363-8_10

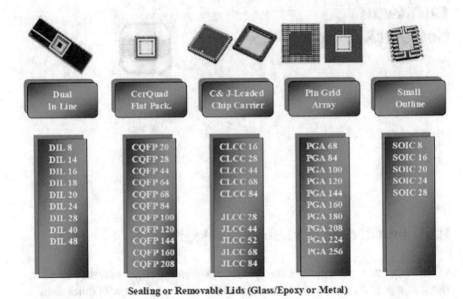

Dual In-Line	CerQuad Flat Pack.	C& J-Leaded Chip Carrier	Pin Grid Array	Small Outline
DIL 8	CQFP 20	CLCC 16	PGA 68	SOIC 8
DIL 14	CQFP 28	CLCC 28	PGA 84	SOIC 16
DIL 16	CQFP 44	CLCC 44	PGA 100	SOIC 20
DIL 18	CQFP 64	CLCC 68	PGA 120	SOIC 24
DIL 20	CQFP 68	CLCC 84	PGA 144	SOIC 28
DIL 24	CQFP 84		PGA 160	
DIL 28	CQFP 100	JLCC 28	PGA 180	
DIL 40	CQFP 120	JLCC 44	PGA 208	
DIL 48	CQFP 144	JLCC 52	PGA 224	
	CQFP 160	JLCC 68	PGA 256	
	CQFP 208	JLCC 84		

Sealing or Removable Lids (Glass/Epoxy or Metal)

Fig. 10.1 Few examples of packages

10.2   Classification of Packages

Depending on the way the leads are arranged on packages and the way they are mounted on the printed circuit boards, there are a variety of package architectures. Based on the arrangement of leads, they are classified as in-line, periphery, and array packages. Based on the way they are mounted, they are classified as through hole or surface mount package architectures. Depending on the material used for the packages, they are classified as plastic or ceramic. Depending on the application and standard to which packaging is manufactured, ceramic packages are classified as military (MIL), automotive, and space, and plastic packages are classified as industry and commercial.

10.3   Criteria for Selection of Packages

Selection of the right package for the SoC depends on criteria listed below:

- Chip performance requirement.
- Power supply IR drop and noise.
- Impedance matching for high-frequency operation.
- Electrical requirements of logic interfaces.
- Chip physical requirement.
- Die size.

- Pin count.
- Thermal requirement.
- Die temperature distribution.
- Package thermal resistance.
- Application environment.
- Hermiticity, temperature, and altitude (SER).
- Form factor.
- Application based, for example, SoCs for smartphones and portable devices.
- Cost.

10.4   Package Components

Typical wire-bonded packages consist of the following parts: planes, bonding wire, and lead planes. The signal from the IO buffer flows through the die pad to a bond wire which lands on the package landing and flows through planes/package routing/ lead frame depending on the type of package and then to the package pin or solder ball (Fig. 10.2).

10.5   Package Assembly Flow

The silicon die is mounted and bonded onto the package base using epoxy or eutec-tic glue, and then each of the die pads is wire bonded to the package landing using wire bonding machine by suitable bonding types like wedge bonding, ball bonding, or ribbon bonding and then it is sealed with lid or mold. The step-by-step flow is shown in Fig. 10.3.

Bond wires are typically made of gold or aluminum of different thickness and it is selected based on the tolerability of parasitic inductance values. The wire bonding process is based on the ultrasonic welding technique or thermo-sonic technique.

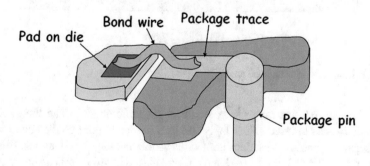

Fig. 10.2 Parts of wire-bonded package

Pin-Grid Array Assembly Process

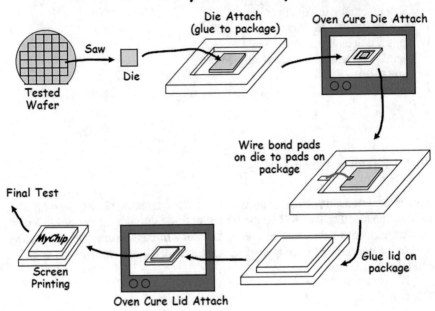

Fig. 10.3 Package assembly flow

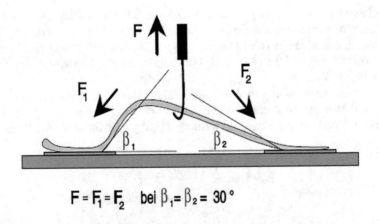

$$F = F_1 = F_2 \quad \text{bei } \beta_1 = \beta_2 = 30°$$

Fig. 10.4 Reliability tests on bond wire

Both wires are bonded on pads as small as 10sq.micron size. The thermo-sonic technique is used to bond solder balls and uses hardened pure gold as bond wires, and ultrasonic bonding uses aluminum wires for high voltage applications. Bonding types can be in-line where the package pins are placed in order or it can be staggered where bond pads are placed in a cross fashion to accommodate more input-outputs.

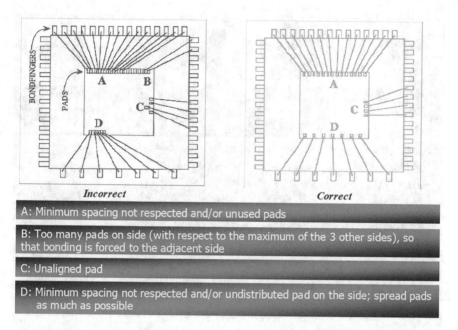

Fig. 10.5 Bonding rules

Quality of bonding is tested by visual inspection using a scanning electron microscope (SEM) and pull and shear tests as shown in Fig. 10.4.

Wire bonding and assembly procedure have to follow bonding rules like physical spacing, length of bond wires, etc. A few examples are shown in Fig. 10.5.

10.6 Packaging Technology

There are many types of packages used in packaging the systems on chip. They are the following:

1. Wire bonded: QFP, BGA, uSTARBGA, etc. (ceramic and plastic) are examples of wire-bonded packages, Few wire-bonded packages are shown in Fig. 10.6.

2. Flip-chip packages: Few examples are FBGA (ceramic and plastic). In this the die is directly flipped and connected to the interconnect patterns on the package substrate through solder balls as shown in Fig. 10.7.

3. Advanced packages with examples such as system in package (SIP) and chip scale package (CSP)/wafer scale packages (WSP). Figure 10.8 shows Pentium Pro SIP package. Figure 10.9 shows wafer scale package from Texas Instruments.

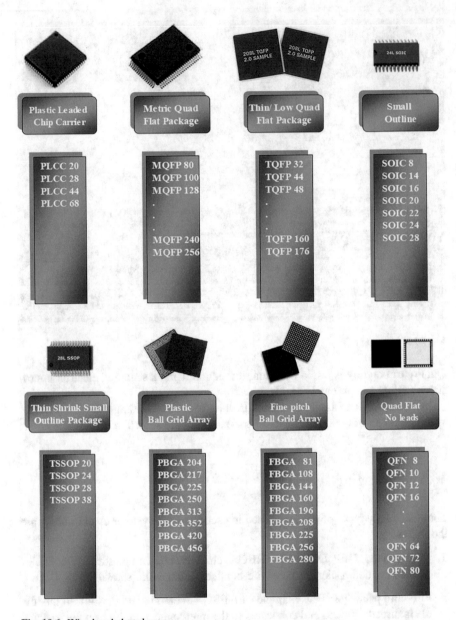

Fig. 10.6 Wire-bonded packages

10.7 Flip-Chip Packages

Flip-chip packages are gaining popularity as they allow for smaller size and pitch and large IO pins and high heat dissipation advantage. In this, the die is directly flipped onto the package which has solder balls routed to the landing.

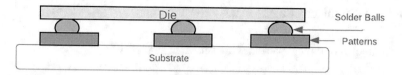

Fig. 10.7 Flip-chip Bonding

Fig. 10.8 Pentium Pro chip package. (Source https://de.wikipedia.org/. Courtesy: Intel)

Fig. 10.9 Wafer chip scale packaging. (Credit by © Raimond Spekking/CC BY-SA 4.0 (via Wikimedia Commons), CC BY-SA 4.0, https://commons.wikimedia.org/w/index.php?curid= 64189136 Source Texas Instruments)

10.8   Typical Packages

Few examples of typical packages are shown in Fig. 10.10.

10.9   Package Performance

Package performance is measured by the electrical tests and mechanical tests performed on them. Electrical tests include tests for pin parasitic effect and simultaneous output switching noise and mechanical tests include heat radiation using thermal models.

10.10   System Integration

Developing system on chip is one aspect of it but packaging is much more advanced in housing many chips in a single "systems in packages" (SIP), where the multiple chips are either wire bonded to each other or flip-chipped. Also the passives, small circuits, SMD devices, and bare dies are all packaged together into one. A few examples of this are shown in Fig. 10.11.

10.11   Packaging Trends

VLSI technology has piggybacked on packaging technology for meeting a never-ending demand for integration of more and more functionalities as multi-chip system solutions. This also has kept alive the "More than Moore" vision. Implementing more complex multifunction system on a chip as a monolithic IC turns out to be super expensive and is not commercially viable. This was predicted by Gordon Moore who stated that "It may prove to be more economical to build large systems out of smaller functions, which are separately packaged and interconnected." Technology like the system in package (SIP) for such systems has worked out well as an alternatives to designing an advanced-node monolithic SoC for heterogeneous ICS. Exploring such alternatives has continued in parallel with Moore's miniaturization over the decades. The evolution of packaging technologies for such integration is shown in Fig. 10.12.

With the growing need for the integration of much more heterogeneous dies-based architectures, new semiconductor package design methodologies are required. This demand arises in advanced nodes less than 7 nm; the interconnect in traditional packages is not sufficient to meet PPA goals. Processing in sub-ten nanometer technologies also introduces new difficulties, to realize nonplanar FinFET transistor

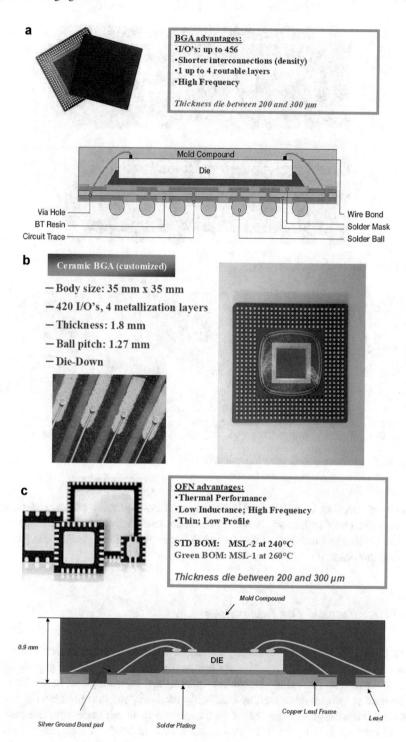

Fig. 10.10 (**a**) BGA package, (**b**) Ceramic BGA. (**c**) QFN package

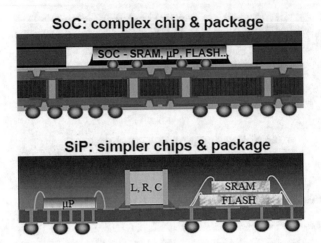

Fig. 10.11 Multi-chip in single package

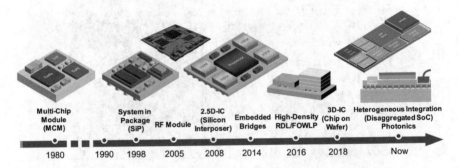

Fig. 10.12 Evolution of packaging technologies

structures, needing complex double-patterning lithography for critical layers, and perhaps even new substrates. Newer systems demand the integration of chips realized in diverse technologies. This has led to the exploration of stacked die integration technologies.

10.11.1 Stacked Die Integration

Over the years, designers constantly strived to catch up with processing technologies to make Moore's law true: first to bridge the design-productivity gap and second to overtake it using innovative packaging technologies. Innovation in interconnect technologies such as multi-chip modules, silicon in the package, and package-on-package schemes has played a major role in this endeavor. The current 3D IC concept is believed to be promising profound levels of integration as it is

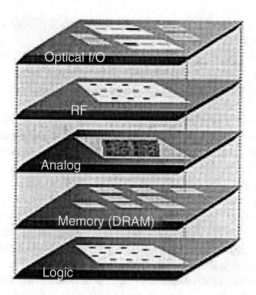

Fig. 10.13 3D IC structure

technology to stack multiple dies through a layer of the interposer. An interposer is a silicon layer used as a bridge or a conduit to allow electrical signals to pass through it from one die to another die or die to board with varying signal pitches. A typical 3D IC structure is shown in Fig. 10.13.

10.11.2 3D Integration Schemes

A few commonly used schemes for the integration of 3D ICs are as follows.

2.5D IC: In this integration concept, two or more dice are placed adjacent to each other facing downward toward the silicon interposer. Dies are supported by micro-bumps on their active surfaces which get connected to the silicon inter-poser. Sometimes, the redistribution layer (RDL) is added to align the micro-bumps to the interposer pads, or signals are routed on the RDL to connect them. Connection from RDL to interposer is achieved through silicon vias (TSVs). There is a possibility of multiple RDLs to the package for complex integration needs.

3D IC: In the 3D IC concept, dies are stacked one above the other and they are interconnected vertically through silicon vias (TSVs). The stacked dies can be similar as in memory blocks or they can be of different functionalities realized using different compatible technologies. The 3D integration of similar dies is called homogeneous 3D technologies and that of dissimilar dies is called heterogeneous 3D technology. 3D heterogeneous integration technology is very complex and most promising in terms of possible levels of integration. Technology offers different approaches to integrating multiple dies.

Face-face: Dies are connected using micro-bumps wherein the lower die also has TSV through active layers and substrate to metallization on its back surface. The back surface of the die acts as RDL using pads onto which pads are laid to connect them to the package substrate.

Face to back: Two or more dies are placed one above the other and connections are made through TSV and DSL and micro-bumps and to the package substrate.

5.5D IC: in this integration approach, the 3D-IC integration is connected to 2.5D-IC silicon interposer to enhance further integration to develop high bandwidth, compute-intensive solutions

Chiplets: Chiplet is a fully functional system component with an interface that is designed to work with other chiplets to form a system on the chip. Integration of chiplets uses advanced packaging technology information during chip design. The concept of multiple chips in a single package is an old technique. Multi-chip modules (MCMs) and system in package (SIP) existed in semiconductor technologies as early as 1980 as shown in Fig. 10.12 Assembling chiplets side by side on the same substrate and interconnecting them is termed as 2.5D technology. Applications like high-performance computing, AI, and systems with incredibly large memories of high-bandwidth and low-latency demand integration of multi-die fabricated by different technologies like MEMS (sensors), RFCMOS, analog, and many others are driving factors for multi-die integration. This has forced designers to look for advancements beyond 2.5D technologies to 3D IC technologies and heterogeneous integration.

3D IC constituents:
The following are different constituents of 3D IC:

- Bumps and balls.
- Ball grid arrays.
- C4 (controlled collapse chip connection) bumps.
- Micro-bumps.
- Through silicon vias.
- Redistribution layers.
- Silicon interposer.

Bumps and balls: Bumps and balls serve to match the interconnect spacings and dimensions to enable the connection between two technologies. The dies are connected to PCBs through these bumps and balls.

Ball grid arrays: A package-level interconnect that connects a packaged device to a PCB.

C4 (controlled collapse chip connection) bumps: Solder balls arranged as an array of grids used to connect bare dies to PCBs. These bumps have a pitch of 180um.

Micro-bumps: Micro-bumps are small solder balls of pitch less than 10 μm that are used to connect two dies face to face.

Through silicon vias (TSV): TSVs use the CMOS etching process to create vias from the active to top of the die and through the backside of the substrate. They are further filled with copper or tungsten to make interconnections from the circuit to the top of the die to the backside. The typical diameter of the TSV is 12um with a 180um pitch. Filling these long vias is very challenging and time consuming. Hence, the wafers are trimmed to 50um thickness from 300 to 350um original thickness. Another main challenge is that these vias travel through active layers of the wafer and the designers must take care of the ESD issues that originate from them during the physical design of the chip planned for 3D with TSV. TSV can also result in mechanical stress which also needs to be addressed.

A redistribution layer (RDL): The redistribution layer shown in Fig. 10.14 consists of metallization on the surface of a die, either on its active face or on the back of the substrate, which is then patterned to redistribute connections from one part of the die to another, or to match the pitch of two interconnection technologies. Redistribution layers have landing bumps on which micro-bumps are formed to make connections.

Silicon interposer: An interposer is a silicon layer used as a bridge or a conduit to allow electrical signals to pass through it from one die to another die or die to board with varying signal pitches. The silicon interposer is shown in Fig. 10.15.

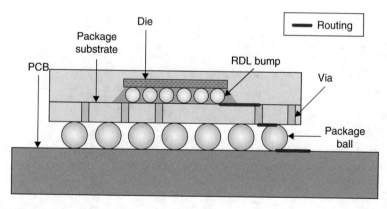

Fig. 10.14 Redistribution layer (RDL). (Source: EDN)

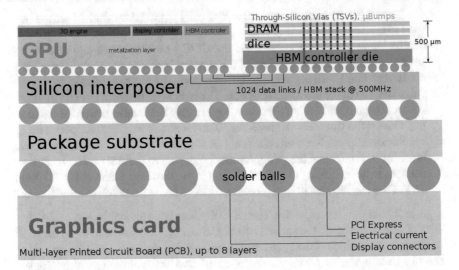

Fig. 10.15 Silicon interposer. (Source: Wikipedia)

Chapter 11
Reference Designs

11.1 Design for Trial

The design examples and case studies presented here can be copied onto the Linux-based design working directory for practice with EDA tools. Design examples and test benches are modelled in Verilog HDL. The simulations can be run, and the results can be compared with the sample waveforms provided against each of the designs in this chapter. These designs can be reused to build larger systems.

11.2 Prerequisites

The user should have working knowledge of Linux commands and any of the text editors, such as Vi editor. For running simulations and debugging, simulator and waveform viewer tools are required. The design examples in Part 1 can be simulated. For the design flow in Part 2 involving synthesis and logic equivalence check (LEC), standard cell library files are required. You also need a synthesis tool for this part. For static timing analysis (STA) and physical design, you require place and route (physical design) tools, a technology library with physical design views, and delay models of the design elements. Therefore, for this part, the scope of reference design is restricted to explanations and indicative scripts using a dummy technology library. Also, an attempt is made to present a near real-time design environment.

© The Author(s), under exclusive license to Springer Nature Switzerland AG 2022
V. S. Chakravarthi, *A Practical Approach to VLSI System on Chip (SoC) Design*,
https://doi.org/10.1007/978-3-031-18363-8_11

11.3   User Guidelines

The database is to be copied to the working directory. The directory structure shown in the next section is with reference to the current working directory. Each of the design directories has a "document file," which will explain the design and the test bench in brief. User can experiment on any standard simulators like NCSim, QuestaSim, and VCS. For running the simulations and particular tool commands, the user is advised to refer to the tool's user manual.

11.4   Design Directory

The directory structure is as shown below:

```
Pwd>://ReferenceDesigns/Examples/adder/design.v
/tb.v
/doc
/Multiplier/ design.v
/tb.v
/run.f
/doc
/Counter design.v
/tb.v
/run.f
/doc
…….. .
/DesignFlow/
. .…….
/Case_study/MBI/
…….. .
/IOT_SOC/
…….. .
```

11.5   Part 1

The following example designs are modeled in Verilog HDL in this section.
   Arithmetic functions:

1. 32-bit adder
2. 16 x 16 multiplier
3. 32-bit counter with overflow
4. 4-bit up/down counter

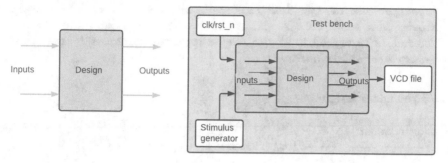

Fig. 11.1 RTL design and test bench structure

Logical function blocks:

5. 2 clients arbiter
6. 8:1 multiplexer
7. 1:8 demultiplexer
8. 4:2 encoder
9. 2:4 decoder
10. 2x2 matrix multiplier
11. 2-bit comparator
12. Finite state machine-based sequence detector (sequence: 10101)
13. Linear feedback shift register (LFSR)
14. Hour-minute-second timer
15. Self-synchronizing scrambler
16. Side stream scrambler-descrambler
17. Colored ball puzzle box
18. Scratchpad register
19. Configuration and status registers
20. Field crossing clocks (clock domain crossover—CDC) block

The design and test bench model are described using behavioral RTL models using Verilog HDL. The RTL and test bench (TB) model structures of the designs are shown in Fig. 11.1.

User can find comments in all the design files within a pair of // which are self-explanatory. The RTL directory in respective design directory has the RTL design file and test_bench (TB) files modeled in Verilog HDL. It also contains sample waveform file which can be used as reference waveforms for simulations. You need waveform viewer tool such as SimVision to open the wave form file in vcd format.

11.6   Design Examples

32-bit adder
Inputs: two 32-bit operands in A and B
    Output: sum_32

*Function: The design adds two operands of 32-bit binary numbers
stored in A and B both 32-bit registers representing the operands.
The result is stored in 33-bit sum_32 register which has {carry, Sum}
Design file: 32bit_adder.v.*

// 32 bit adder design

```
module adder (
//------------------clock_reset------------------//
 clk ,
 reset_n ,
//---------------Input-------------------------//
 en ,
 op_a ,
 op_b ,
//--------------output-------------------------//
 adder_out,
 carry_out
 );
//input-output declaration
input clk , reset_n ;
//---------------Input---------------------//
input en ;
input [31:0] op_a ;
input [31:0] op_b ;
//-------------output---------------------//
output [31:0] adder_out ;
output carry_out ;
//Internal signal declaration

reg [32:0] adder_reg ;


assign adder_out = adder_reg[31:0] ;
assign carry_out = adder_reg[32] ;

always@(posedge clk or negedge reset_n)
begin
 if (!reset_n) begin
 adder_reg<=33'd0;
 end
else begin
 if (en) begin
 adder_reg <=op_a + op_b ; // en is the enable to carry the addi-
```

```
tion of two numbers.
 end
 end
end
endmodule
```

Test bench module adder_tb

Inputs: Nil
Outputs: Nil
Function: The test bench applies random values to A and B operands and checks the result of addition by generating a signal match to indicate the correct behavior. The waveform 32-bit_adder.vcd is written out which can be observed using waveform viewer.

Test bench file: 32-bit_adder_tb.v

```
module adder_tb;
//--------------- Inputs--------
 reg clk;
 reg reset_n;
 reg en;
 reg [31:0] A;
 reg [31:0] B;
//----------------- Outputs-----------
 wire [31:0] sum;
 wire carry_out;
// clock generation
always #5 clk = ~clk; // toggle clock for every 5 ticks
initial begin
 clk = 0;
 reset_n = 1;
 en = 0;
 $display("--------- Test Started ---------");
 #10 reset_n = 0;
 #10 reset_n = 1;
 en = 1;

 $display("--------- Sending Data A = 32'hAAAAAAAA and B =
32'hEEEEEEEE ---------");


 A = 32'hAAAAAAAA;
 B = 32'hEEEEEEEE;
```

```verilog
 $display("--------- Sending Data A = 32'h7777777 and B =
32'h2456321 ---------");


 #10
 A = 32'h7777777;
 B = 32'h2456321;

$display("--------- Sending Data A = 32'hCCCCCCCC and B =
32'hBBBBBBB ---------");
#10
 A = 32'hCCCCCCCC;
 B = 32'hBBBBBBB;

$display("--------- Sending Data A = 32'h11111111 and B =
32'b11111111 ---------");
#10
A = 32'h11111111;
B = 32'h11111111;
$display("--------- Test Ended ---------");
 #1000 $finish;
end

//module instantiation
adder u_adder(
 .clk(clk),
 .reset_n(reset_n),
 .en(en),
 .op_a(A),
 .op_b(B),
 .adder_out(sum),
 .carry_out(carry_out)
 );

initial
 begin
 $dumpfile("adder_tb.vcd");
 $dumpvars(0,adder_tb);
 end
endmodule
```

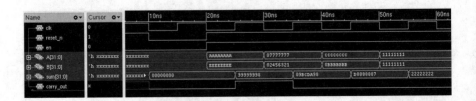

16 x 16 Multiplier

```
16 x 16 multiplier
Inputs: two 16-bit operands in A and B
Outputs: 16-bit multi_out
Function: The design performs multiplication of two operands of
16-bit binary numbers stored in A and B both 16-bit registers rep-
resenting the operands. The result is stored in 16-bit register.
Design file: multiplier.v


/***************************************
// Module works for 16x16 multiplier of A and B.
// This is combinational block which doesn't require clock and
reset //
//User can refer to any Verilog HDL language book to understand the
syntax of commands. //
***************************************/
//16*16 bit multiplier
module multiplier (
//-----------------clock_reset-----------------//
 clk ,
 reset_n ,
//--------------Input--------------------//
 en ,
 op_a ,
 op_b ,
//------------output---------------------//
 multi_out
 );
//----------------clock_reset----------------//
input clk ,
 reset_n ;
//--------------Input--------------------//
input en ;

input [15:0] op_a ,
 op_b ;
//------------output---------------------//
output [31:0] multi_out ;

reg [31:0] multi_out_reg ;


assign multi_out = multi_out_reg ;
```

```verilog
always@(posedge clk or negedge reset_n)
 begin
 if (!reset_n) begin
 multi_out_reg<=32'd0;
 end
 else begin
 if (en)
 multi_out_reg<= (op_a * op_b);
 end
end
endmodule
```

Test bench module multiplier_tb

```verilog
Inputs: Nil
Outputs: Nil
Function: The test bench applies random values of A and B and
result is stored in 16 register. The waveform multiplier_tb.vcd
can be observed using waveform viewer.
/***********************************************************/
Test bench file: multiplier_tb.v
module multiplier_tb;

reg clk;
reg reset_n;
reg en;
reg [15:0] op_a;
reg [15:0]op_b;

wire [31:0] multi_out ;


multiplier u1 (clk,reset_n,en,op_a,op_b,multi_out);
always #5 clk=~clk;

initial
 begin
 clk =0;
 reset_n=0;
 en=0;
 op_a=0;
 op_b=0;

 #10 reset_n=0;
 #10 reset_n=1;
 en =1;
```

```
 op_a = 16'hAAAA;
 op_b = 16'hBBBB;

#10 op_a = 16'h4444;
 op_b = 16'h1111;
#100 $finish;
end

initial
 begin
 $dumpfile("multiplier_tb.vcd");
 $dumpvars(0,multiplier_tb);
 end
endmodule
```

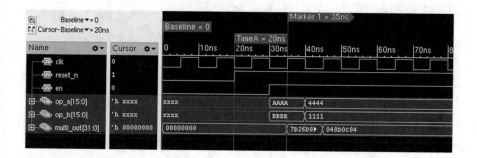

32-bit counter with overflow

```
32-bit counter_overflow
Inputs: en, load
Outputs: counter
Function: The design, when enabled, is high when counter starts
counting, and when the load is made high, 33'hfffffff8 is loaded to
counter_out and the result is stored in register counter_33 {coun-
ter_out_32, counter_overflow}.

Design file: counter_overflow.v
/*************************************
// Module starts 32-bit counting and when load is made high,
33'hfffffff8 is loaded to counter_out.
// This is sequential block which require clock and reset //
//User can refer to any Verilog HDL language book to understand the
syntax of commands. //
*************************************/
//32-bit counter with overflow design
```

```verilog
module counter_overflow(
//-----------------clock_reset-----------------//
 clk ,
 reset_n ,
//--------------Input--------------------//
 en ,
 load,
//-------------output---------------------//
 counter_out ,
 counter_overflow
 );

//Input Output declaration


input clk ,
 reset_n ;
//---------------Input--------------------//
input en ;
input load;
//-------------output--------------------//
output [31:0] counter_out ;

output counter_overflow;


//Internal signal declaration
reg [32:0] counter_reg ;
wire load;

assign counter_overflow= counter_reg[32] ;
assign counter_out = counter_reg[31:0] ;

always@(posedge clk or negedge reset_n)
begin
 if (!reset_n) begin
 counter_reg<=33'd0;

 end
 if(load)
 counter_reg<=33'b111111111111111111111111111111000;
 if (en)
 counter_reg<=counter_reg+33'd1 ;
 end
endmodule
```

Test bench module counter_overflow_tb

```
Inputs: Nil
Outputs: Nil
Function: The test bench applies random values of enable and load
and checks the result of 32-bit counting. The waveform counter_
overflow_tb.vcd is written out which can be observed using wave-
form viewer.
Test bench file: counter_overflow_tb.v
module counter_overflow_tb;

 // Inputs
 reg clk;
 reg reset_n;
 reg en;
 reg load;

 // Outputs
 wire [31:0] counter_out;
 wire counter_overflow;

always
 #5 clk = ~clk;

 initial
 begin
 clk = 0;
 reset_n = 0;
 en = 0;
 load = 0;

 #10 reset_n = 0;
 #10 reset_n = 1;
 #10 en = 1;
 #10 load =1;
 #80 en=0;
#10 en=1;

#10000 $finish;
 end

 counter_overflow uut (
 .clk(clk),
 .reset_n(reset_n),
 .en(en),
```

```
.counter_out(counter_out),
.counter_overflow(counter_overflow),
.load(load)
);
initial
begin
$dumpfile("counter_overflow _tb.vcd");
$dumpvars(0, counter_overflow _tb);
end
endmodule
```

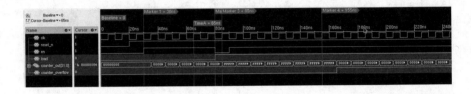

4-bit up/down counter

4-bit up/down counter
Inputs: en
Outputs:up_counter, down_counter
Function: The design, when enabled, is high when the up_counter
starts counting from 0000 to 1111. The down counter starts counting
from 1111 to 0000.

Design file: updowncounter.v

```
/****************************************
// Module starts 4-bit up counting and 4-bit down counting
// This is sequential block which require clock and reset //
//User can refer to any Verilog HDL language book to understand the
syntax of commands. //

****************************************/
//4-bit counter design
module updowncounter(
 clk,
 resetn,
 en,
 up_counter,
 down_counter
 );
```

```
//-----------------input ports------------
input clk;//input clock of the design
input resetn;// avtive low reset
input en;// active high enable
//---------------output ports------------
output[3:0] up_counter;
output[3:0] down_counter;
//---------------input datatype------------
wire clk;
wire resetn;
wire en;
//---------------output datatype------------
reg [3:0] up_counter;
reg[3:0] down_counter;
// for every posedge of the clock below function has to happen
always @(posedge clk or posedge resetn)
begin
if( !resetn)/*if reset is zero, reset upcounter to 0000 down-
counter to 1111*/
begin
up_counter <= 4'b0000;
down_counter <=4'b1111;
end
else if(en)
begin
up_counter <= up_counter + 4'b0001;// incrementing the count value
down_counter<= down_counter-4'b0001;// decrementing the count value
end
end
endmodule
```

Test bench module counter_tb

```
Inputs: Nil
Outputs: Nil
Function: The test bench applies random values and checks the
result of counting. The waveform updown_counter_tb.vcd can be
observed using waveform viewer.
Test bench file: updown_counter_tb.v
module updown_counter_tb;
 // Inputs
 reg clk;
 reg resetn;
 reg en;

 // Outputs
```

```verilog
wire [3:0] up_counter;
wire [3:0] down_counter;

// clock generation
always #5 clk = ~clk; // toggle clock for every 5 ticks

initial begin
// Initialize Inputs
clk = 0;
resetn = 1;
en = 0;

//$display("--------- Test Started ---------");
#10 resetn = 0;
#10 resetn = 1;
en = 1;

#500 $finish;
end

counter uut (
.clk(clk),
.resetn(resetn),
.en(en),
.up_counter(up_counter),
.down_counter(down_counter)
);

initial begin
$dumpfile("updown_counter_tb.vcd");
$dumpvars(0,updown_counter_tb);
end
endmodule
```

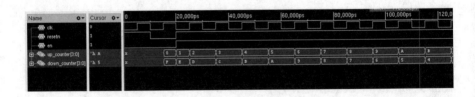

2 clients arbiter

```
Inputs: Request from client 1 and client 2
Outputs: Grant 1, Grant 2
Function: The design grants the request to clients (master) based
on priority. If priority select is high, the request is granted to
client 1 and then the request is granted to client 2.
Design file: arbiter.v
/***************************************
// Module grants request to the respective clients. If both the
clients request at the same time based on the priority the request
is granted to client 1 followed by client 2.
// This is sequential block which require clock and reset //
//User can refer to any Verilog HDL language book to understand the
syntax of commands.
***************************************/
// arbiter design
module arbiter (
//----------------input_data----------------------//
 clk ,
 reset_n ,
//--------------Input_interface--------------------//
 priority_sel , //1- client1 0- client2
 client1_req ,
 client2_req ,
//----------------Output_interface------------------//
 o_grant1 ,
 o_grant2
 );
//----------------input_data----------------------//
input clk ,
 reset_n ;
//--------------Input_interface--------------------//
input priority_sel , //0- client1 1- client2
 client1_req ,
 client2_req ;
//----------------Output_interface------------------//
output o_grant1 ,
 o_grant2 ;
reg [1:0] curr_state ,
 next_state ;
reg client1_req_d ,
 client2_req_d ;
parameter IDLE = 2'd0 ,
 CLIENT1 = 2'd1 ,
```

```verilog
CLIENT2 = 2'd2 ;

always@( client1_req_d ,
 client2_req_d ,
 curr_state ,
 priority_sel
 )
begin
 case (curr_state)

 IDLE : if (priority_sel && client1_req_d)
 next_state = CLIENT1 ;
 else if (client2_req_d)
 next_state = CLIENT2 ;
 else
 next_state = IDLE ;

 CLIENT1 : if ( client2_req_d )
 next_state = CLIENT2 ;
 else
 next_state = IDLE ;

 CLIENT2 : if ( client1_req_d )
 next_state = CLIENT1 ;
 else
 next_state = IDLE ;

 default : next_state = IDLE ;


 endcase
end
always@(posedge clk or negedge reset_n)
begin
 if (!reset_n ) begin
 curr_state<=2'd0;
 end
 else begin
 curr_state<=next_state ;
 end
end
assign o_grant1 = (curr_state == CLIENT1 ) ;
assign o_grant2 = (curr_state == CLIENT2 ) ;

always@(posedge clk or negedge reset_n)
```

```
begin
 if (!reset_n ) begin
 client1_req_d<=1'd0;
 client2_req_d<=1'd0;
 end
else begin
 if (o_grant1)
 client1_req_d<=1'd0;
 else if (client1_req)
 client1_req_d <=1'd1;
 if (o_grant2)
 client2_req_d<=1'd0;
 else if (client2_req)
 client2_req_d <=1'd1;
 end
end
endmodule
```

Test bench module arbiter_tb

```
Inputs: Nil
Outputs: Nil
Function: The test bench applies random requests from client 1 and
client 2 and checks the result of granting the request. The wave-
form arbiter_tb.vcd can be observed using waveform viewer.

Test bench file: arbiter_tb. v
module arbiter_tb;

 // Inputs
 reg clk;
 reg reset_n;
 reg priority_sel;
 reg client1_req;
 reg client2_req;

 // Outputs
 wire o_grant1;
 wire o_grant2;

 initial begin
 clk=1'd0;
 forever #5 clk=~clk;
 end
```

```verilog
arbiter uut (
.clk(clk),
.reset_n(reset_n),
.priority_sel(priority_sel),
.client1_req(client1_req),
.client2_req(client2_req),
.o_grant1(o_grant1),
.o_grant2(o_grant2)
);

initial begin
clk = 0;
reset_n = 0;
priority_sel = 0;
client1_req = 0;
client2_req = 0;
end

initial begin
#10 reset_n =0;
#10 reset_n = 1;
@(posedge clk)
#10 priority_sel = 1;

client1_req = 1;
client2_req = 0;

#10
client1_req = 0;
client2_req = 1;

#10
client1_req = 0;
client2_req = 0;

#10
priority_sel = 0;
client1_req = 1;
client2_req = 1;

#10
priority_sel = 1;
client1_req = 1;
client2_req = 1;
#100 $finish;
```

```
 end
initial begin
$dumpfile("arbiter_tb.vcd");
$dumpvars(0,arbiter_tb);
end
endmodule
```

8:1 multiplexer

```
Inputs: din_8, select lines_3
Outputs: dout
Function: The design works based on the select lines, and appropri-
ate output for given input is generated.
Design file: mux8x1.v
/**************************************
// Module works based on the select lines. If select line is 1 1st
input is selected and goes on..
// Demux is a combinational block which doesn't require clock and
reset but the output from
// demux is latched on clokedge as can be seen in the model. //
// User can refer to any Verilog HDL language book to understand
the syntax of commands. //
**************************************/
//8:1 multiplexer
module mux8x1(
 clk,// clock input of the design
 rstn,// avtive low reset
 en,// avtive high enable
 din, //data input
 sel,// select lines
 dout// data output
 );
```

```verilog
//--------------------------input port------------
input clk;
input rstn;
input en;
input [7:0] din;
input [2:0] sel;
//-------------------------output port-------------
output dout;

//--------------------------input datatype=---------
wire clk;
wire rstn;
wire en;
wire [7:0] din;
wire [2:0] sel;
 // ---------------------output datatype--------------
reg dout;
// for every posedge of the clock below operation should take place
always @(posedge clk or negedge rstn)
begin
if (!rstn)
dout = 0;
else if (en)
case(sel)
3'b000:dout=din[0];
3'b001:dout=din[1];
3'b010:dout=din[2];
3'b011:dout=din[3];
3'b100:dout=din[4];
3'b101:dout=din[5];
3'b110:dout=din[6];
3'b111:dout=din[7];
endcase
end
endmodule
```

Test bench module mux8x1_tb

```verilog
Inputs: Nil
Outputs: Nil
Function: The test bench applies random values to 3-bit select
lines and check the dout. The waveform mux8x1_tb.vcd can be observed
using waveform viewer.
Test bench file: mux8x1_tb.v
module mux8x1_tb;
```

```verilog
// Inputs
reg clk;
reg rstn;
reg en;
reg [7:0] din;
reg [2:0] sel;

// Outputs
wire dout;

// clock generation
always #5 clk = ~clk; // toggle clock for every 5 ticks

initial begin
// Initialize Inputs
clk = 0;
rstn = 1;
en = 0;
//$display("--------- Test Started ---------");
#10 rstn = 0;
#10 rstn = 1;
en = 1;

sel=3'b000; din = 8'b00000001;
#10 sel=3'b001; din = 8'b00000010;
#10 sel=3'b010; din = 8'b00000100;
#10 sel=3'b011; din = 8'b00001000;
#10 sel=3'b100; din = 8'b00010000;
#10 sel=3'b101; din = 8'b00100000;
#10 sel=3'b110; din = 8'b01000000;
#10 sel=3'b111; din = 8'b10000000;
#10 sel=3'b111; din = 8'b00000000;
#10 sel=3'b110; din = 8'b10000000;
#10 sel=3'b100; din = 8'b00010000;
#100 $finish;
end
mux8x1 uut (
.clk(clk),
.rstn(rstn),
.en(en),
.din(din),
.sel(sel),
.dout(dout)
);
initial
```

```
begin
$dumpfile("mux8x1_tb.vcd");
$dumpvars(0,mux8x1_tb);
end
endmodule
```

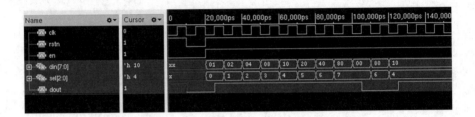

1:8 demultiplexer

```
Inputs: din_8, select lines_3
Outputs: dout
Function: The design works based on the select lines, and appropri-
ate output for given input is generated.
Design file: demux1x8.v
/***************************************
// Module works based on the select lines. If select lines is 2,
the 2nd bit in output will be high //
// and rest will be zeros.
// This is combinational block which doesn't require clock and
reset but the //
// output is latched using clock. //
// User can refer to any Verilog HDL language book to understand
the syntax of commands. //
****************************************/
//1:8 demultiplxer with 3 selectlines

module demux1x8(
  clk,
  rstn,
  en,
  sel,
  din,
  dout
  );
```

```
//-------------input ports-----------
input clk; // input clock of the design
input rstn;// active low reset
input en;// active high enable
input [2:0] sel;// select lines
input din;// datain

//-------------output ports-----------


output [7:0] dout;// output data


//-------------input datatypes-----------


wire clk;
wire rstn;
wire en;
wire din;
wire [2:0] sel;
//-------------output datatypes-----------

reg [7:0] dout;


// for every postitive edge of clock perform below operation
always @(posedge clk or negedge rstn)
begin
if (!rstn) // check condition reset=0,reset dout to 0
dout = 0;
else if (en)
case(sel)
3'b000:begin
 dout[0]=din;
 dout[7:1]=7'b0;
 end
3'b001:begin
 dout[1]=din;
 dout[0]=1'b0;
 dout[7:2]=6'b0;
 end
3'b010:begin
 dout[2]=din;
 dout[1:0]=2'b0;
```

```
  dout[7:3]=5'b0;
  end
3'b011:begin
  dout[3]=din;
  dout[2:0]=3'b0;
  dout[7:4]=4'b0;
  end
3'b100:begin
  dout[4]=din;
  dout[3:0]=4'b0;
  dout[7:5]=3'b0;
  end
3'b101:begin
  dout[5]=din;
  dout[4:0]=5'b0;
  dout[7:6]=2'b0;
  end
3'b110:begin
  dout[6]=din;
  dout[5:0]=6'b0;
  dout[7]=1'b0;
  end
3'b111:begin
  dout[7]=din;
  dout[6:0]=7'b0;
  end
endcase
end

endmodule
```

Test bench module demux1x8_tb

```
Inputs: Nil
Outputs: Nil
Function: The test bench applies random values to 3-bit select
lines and check the dout. The waveform demux_tb.vcd can be observed
using waveform viewer.
Test bench file: demux1x8_tb.v
module demux1x8_tb;

  // Inputs
  reg clk;
  reg rstn;
```

```verilog
reg en;
reg [2:0] sel;
reg din;

// Outputs
wire [7:0] dout;

// clock generation
always #5 clk = ~clk; // toggle clock for every 5 ticks

initial begin
 clk = 0;
 rstn = 0;
 en = 0;

 //$display("--------- Test Started ---------");
 #10 rstn = 0;
 #10 rstn = 1;
 en = 1;

 sel=3'b000; din = 1'b1;
 #10 sel=3'b001; din = 1'b1;
 #10 sel=3'b010; din = 1'b1;
 #10 sel=3'b011; din = 1'b1;
 #10 sel=3'b100; din = 1'b1;
 #10 sel=3'b101; din = 1'b1;
 #10 sel=3'b110; din = 1'b1;
 #10 sel=3'b111; din = 1'b1;

 #100 $finish;
end

 demux1x8 uut (
 .clk(clk),
 .rstn(rstn),
 .en(en),
 .sel(sel),
 .din(din),
 .dout(dout)
 );
initial
 begin
 $dumpfile("demux1x8_tb.vcd");
 $dumpvars(0,demux1x8_tb);
 end
```

endmodule

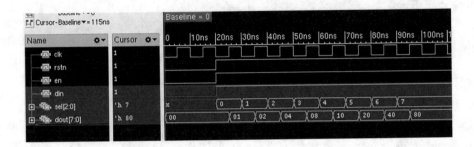

4:2 encoder

Inputs: 4-bit din
Outputs: 2-bit dout
Function: The design encodes 4-bit din
Design file: encoder4x2.v
/**************************************
// Module starts encoding 4-bit din
// This is combinational block which doesn't require clock and reset. But clock used to latch the output.//
//User can refer to any Verilog HDL language book to understand the syntax of commands. //
**************************************/
//4:2 encoder
module encoder4x2(
 din,clk,
 dout,rstn,
 en
 );
//-----------------------input ports------------
input en;// active high enable
 input clk;// clock input of the design
input rstn;// avtive low reset
input [3:0]din;// 4 bit input data

//-----------------------output ports------------

output [1:0] dout;//2 bit output data

//-----------------------input datatypes------------

```
wire en;
wire rstn;
wire [3:0]din;
//-----------------------output datatypes------------

reg [1:0]dout;


// for every positive edge of the clock below operation has to
take place
always @( posedge clk or negedge rstn)
begin
if(!rstn)
dout=2'b00;
else if(en)

case(din)
4'b0001:dout=2'b00;
4'b0010:dout=2'b01;
4'b0100:dout=2'b10;
4'b1000:dout=2'b11;
default dout=2'b00;
endcase
end
endmodule
```

Test bench module encoder4x2_tb

```
Inputs: Nil
Outputs: Nil
Function: The test bench applies random values to 4-bit din and
check the encoded 2-bit dout. The waveform encoder4x2_tb.vcd can
be observed using waveform viewer.
Test bench file: encoder4x2_tb.v

module encoder4x2_tb;


// Inputs
reg [3:0] din;
reg en;
reg clk;
reg rstn;
```

```verilog
 // Outputs
 wire [1:0] dout;
// clock generation
always #5 clk = ~clk; // toggle clock for every 5 ticks

 initial begin
 // Initialize Inputs
 clk = 0;
 rstn = 1;
 en = 0;

//$display("--------- Test Started ---------");
 #10 rstn = 0;
 #10 rstn = 1;
 en = 1;

 din = 4'b0001;
 #10 din = 4'b0010;
 #10 din = 4'b0100;
 #10 din = 4'b1000;

#100 $finish;
end
 encoder4x2 uut (
 .clk(clk),
 .din(din),
 .dout(dout),
 .rstn(rstn),
 .en(en)
 );

 initial
 begin
 $dumpfile("encoder4x2_tb.vcd");
 $dumpvars(0,encoder4x2_tb);
 end
endmodule
```

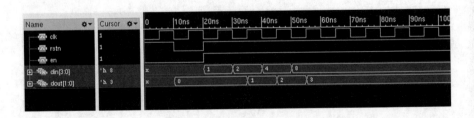

2:4 decoder

```
Inputs: 2-bit din
Outputs: 4-bit dout
Function: The design decodes 2-bit din.
Design file: decoder2x4.v
/***************************************
// Module starts decoding 2-bit din
// This is combinational block which doesn't require clock and
reset, but used //
// to latch the output //
//User can refer to any Verilog HDL language book to understand the
syntax of commands. //
***************************************/
//2:4 decoder
module decoder2x4(
 clk,
 rstn,
 en,
 din,
 dout
 );

//--------------input ports---------------
input en;// active high enable
input clk;// input clock of the design
input rstn;// active low reset
input [1:0]din;// input data

//--------------output ports---------------
output [3:0]dout;// output data
//--------------input datatypes---------------
wire clk;
wire en;
wire rstn;
wire [1:0]din;
//--------------output datatypes ports---------------
reg [3:0]dout;
// for every positive edge of the clock below operation take place
always @( posedge clk or negedge rstn)
begin
if(!rstn)// check condition reset=0, reset the dout to 0
dout=4'b0000;
else if(en)
case(din)
```

```
2'b00:dout=4'b0001;
2'b01:dout=4'b0010;
2'b10:dout=4'b0100;
2'b11:dout=4'b1000;
default dout=4'b0000;
endcase
end
endmodule
```

Test bench module decoder2x4_tb

```
Inputs: Nil
Outputs: Nil
Function: The test bench applies random values to 2-bit din and
check the decoded 4-bit dout. The waveform decoder_tb.vcd can be
observed using waveform viewer.
Test bench file: decoder2x4_tb.v
module decoder2x4_tb;

  // Inputs
  reg clk;
  reg rstn;
  reg en;
  reg [1:0] din;

  // Outputs
  wire [3:0] dout;
  // clock generation
always #5 clk = ~clk; // toggle clock for every 5 ticks

  initial begin
  // Initialize Inputs
  clk = 0;
  rstn = 1;
  en = 0;

//$display("--------- Test Started ---------");
  #10 rstn = 0;
  #10 rstn = 1;
  en = 1;

  din = 2'b00;
  #10 din = 2'b01;
  #10 din = 2'b10;
  #10 din = 2'b11;
  #100 $finish;
```

```
end
decoder uut (
.clk(clk),
.rstn(rstn),
.en(en),
.din(din),
.dout(dout)
);

initial
begin
$dumpfile("decoder2x4_tb.vcd");
$dumpvars(0,decoder2x4_tb);
end
endmodule
```

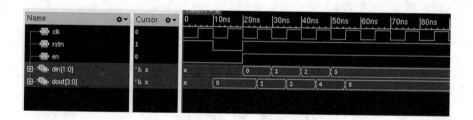

2x2 Matrix Multiplication

```
2x2 matrix multiplication
Inputs: two 32-bit operands in A and B
Outputs: Res;
Function: The design performs matrix multiplication of two oper-
ands of 32-bit binary numbers stored in A and B both 32-bit regis-
ters representing the operands. The result is stored in 32-bit
Res_32 register.
Design file: matrix2x2_mult.v
/****************************************
// Module works for 2x2 matrix multiplication. Both the inputs are
converted to 1D to 3D //
// array and becomes and each rows and columns will have 8 bit. //
// This is combinational block which doesn't require clock and
reset //
//User can refer to any Verilog HDL language book to understand the
syntax of commands. //
****************************************/
//2x2 matrix multiplication
module matrix2x2_mult(A, B, Res, clk, rstn, en);
```

```verilog
//-------------input port--------------------
 input clk, rstn, en;
 input [31:0] A;
 input [31:0] B;
// -----------------------output port-----------
 output [31:0] Res;

//-----------------input datatype------------
 wire clk,rstn,en;
//-----------------output datatype------------
 reg [31:0] Res;
 reg [7:0] A1 [0:1][0:1];
 reg [7:0] B1 [0:1][0:1];
 reg [7:0] Res1 [0:1][0:1];

//for ever A and B value below format should be adopted
always@ ( A or B )
 begin
 {A1[0][0],A1[0][1],A1[1][0],A1[1][1]} = A;
 {B1[0][0],B1[0][1],B1[1][0],B1[1][1]} = B;
 end
//for every posedge of clock below operation should take place

 always@ ( posedge clk or negedge rstn )
 begin
 if(!rstn) begin
 {Res1[0][0],Res1[0][1],Res1[1][0],Res1[1][1]} = 32'd0;
 end
 else
 if(en) begin
 Res1[0][0] =(A1[0][0]*B1[0][0]) + (A1[0][1]*B1[1][0]);
 Res1[0][1] =(A1[0][0]*B1[0][1]) + (A1[0][1]*B1[1][1]);
 Res1[1][0] =(A1[1][0]*B1[0][0]) + (A1[1][1]*B1[1][0]);
 Res1[1][1] =(A1[1][0]*B1[0][1]) + (A1[1][1]*B1[1][1]);

 Res = {Res1[0][0],Res1[0][1],Res1[1][0],Res1[1][1]};
 end
end
endmodule
```

Test bench module matrix2x2_mult_tb

Inputs: Nil
Outputs: Nil
Function: The test bench applies random values of A and B and

result is stored in 32-bit Res. The waveform matrix2x2_mult_tb.vcd
can be observed using waveform viewer.
Test bench file: matrix2x2_mult_tb.v

```verilog
module matrix2x2_tb();
 reg [31:0] A;
 reg [31:0] B;
 reg clk;
 reg rstn;
 reg en;
// Outputs
 wire [31:0] Res;
 always #5 clk = ~clk;
 initial begin
 clk =0;
 rstn =0;
 en =0;
 A = 0;
 B = 0;

 #10 rstn =0;
 #10 rstn =1;
 #10 en =1;
 A=32'b00000001000000010000000100000001;
 #10 B=32'b00000001000000010000000100000001;

 #10 A=32'b00000010000000010000000100000010;
 #10 B=32'b00000010000000010000000100000010;
#100 $finish;
 end
 matrix2x2_mult uut (
 .A(A),
 .B(B),
 .Res(Res),
 .clk(clk),
 .rstn(rstn),
 .en(en)
 );
 initial begin
 $dumpfile("matrix2x2_mult_tb.vcd");
 $dumpvars(0, matrix2x2_mult_tb);
 end
endmodule
```

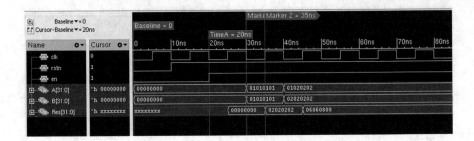

2-bit comparator

2-bit comparator
Inputs: A and B
Outputs: a-greater than-b, a-equal to-b, a-lesser than-b
Function: The design compares inputs A and B. If A is greater than
B, the result is stored in a_grtr_b. If A is lesser than B, the
result is stored in a_lesr_b. If A is equal to B, the result is
stored in a_eql_b.

Design file: comparator.v
/**
// Module compares the 2-bit input A and B and gives the result
whether A is greater than b //
//or A lesser than B or A equal to B.This is combinational block
which doesn't require //
//clock and reset. User can refer to any Verilog HDL language book
to understand the //
//syntax of commands. //
***/
// Comparator design
module comparator (
 clk,
 rstn,
 en,
 A,
 B,
 a_grtr_b,
 a_lsr_b,
 a_eql_b
 );
//----------------input ports--------------
 input clk;// input clock of the design
 input rstn;// active low reset
 input en;// active high enable
 input [1:0] A;
 input [1:0] B;

```verilog
//----------------output ports--------
 output a_grtr_b;
 output a_lsr_b;
 output a_eql_b;
//----------------input datatype----------
 wire clk;
 wire rstn;
 wire en;
 wire [1:0]A;
 wire [1:0] B;
//----------------output datatype---------------
 reg a_grtr_b;
 reg a_lsr_b;
 reg a_eql_b;
// at every posedge of the clock
 always@(posedge clk or negedge rstn)
begin
 if(!rstn)// reset all the values to zero if rstn is 0
 begin
 a_grtr_b = 1'b0;
 a_lsr_b = 1'b0;
 a_eql_b = 1'b0;
 end
 else if (en)// if enable is high start comparing the inputs
 begin
  a_grtr_b  =   ((A[1]&(~B[1]))|(A[0]&(~B[0])&(~B[1]))|(A[0]&A[1]&
(~B[0]))));
   a_lsr_b   =    (((~A[1])&B[1])|((~A[0])&A[1]&B[1])|((~A[1])&B[0
]&B[1]));
 a_eql_b =(((~A[0])&(~A[1])&(~B[0])&(~B[1]))|((A[0]&(~B[0])&(~B[
1]))|(A[0]&A[1]&B[0]&B[1])|((~A[0])&A[1]&(~B[0])&B[1])));
 end
 end
endmodule
```

Test bench module comparator_tb

```
Inputs: Nil
Outputs: Nil
Function: The test bench applies random values to A and B and
checks the results of comparison between them. The waveform com-
partor_tb.vcd can be observed using waveform viewer.

Test bench file: comparator_tb
module comparator_tb;
```

```
// Inputs
reg clk;
reg rstn;
reg en;
reg [1:0] A;
reg [1:0] B;

// Outputs
wire a_grtr_b;
wire a_lsr_b;
wire a_eql_b;

// clock generation
always #5 clk = ~clk; // toggle clock for every 5 ticks

initial begin
// Initialize Inputs
clk = 0;
rstn = 1;
en = 0;
A = 0;
B = 0;

//$display("--------- Test Started ---------");


#10 rstn = 0;
#10 rstn = 1;
en = 1;


A=2'b00;B=2'b00;
#10 A=2'b01;B=2'b10;
#10 A=2'b10;B=2'b00;
#10 A=2'b11;B=2'b11;
#10 A=2'b10;B=2'b01;

#100 $finish;
end

comparator uut (
.clk(clk),
.rstn(rstn),
.en(en),
.A(A),
.B(B),
```

```
.a_grtr_b(a_grtr_b),
.a_lsr_b(a_lsr_b),
.a_eql_b(a_eql_b)
);
initial
begin
$dumpfile("comparator_tb.vcd");
$dumpvars(0,comparator_tb);
end

endmodule
```

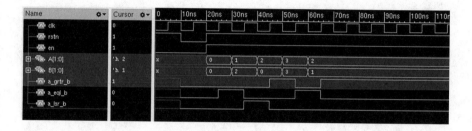

Finite state machine-based sequence detector (pattern: 10101)

```
Sequence detector of 10101 without overlap
Inputs: serial input data
Outputs: seq_detected
Function: The design works to detect the sequence 10101 for which
the output seq_detected will be high.
Design file: fsm.v
/****************************************
// Module works only to detect the pattern 10101
// This is sequential block which require clock and reset //
//User can refer to any Verilog HDL language book to understand the
syntax of commands. //
****************************************/
// Sequence detector of 10101 without overlap
module fsm (
//-----------------clock_reset-----------------//
 clk ,
 reset_n ,
//---------------Input--------------------//
 input_data ,
//-------------Output--------------------//
 seq_detected
 );
```

```
//----------------clock_reset-----------------//
input clk ,
 reset_n ;
//---------------Input--------------------//
input input_data ;
//-------------Output---------------------//
output seq_detected ;

reg [2:0] curr_state ,
 next_state ;

parameter IDLE =3'd0 ,
 SEQ_A =3'd1 ,
 SEQ_B =3'd2 ,
 SEQ_C =3'd3 ,
 SEQ_D =3'd4 ;

//-----------------next_state_logic------------------------
-------//
always@ ( curr_state ,
 input_data
 )
begin
 case (curr_state)

 IDLE : if (input_data)
 next_state= SEQ_A ;
 else
 next_state= IDLE;

 SEQ_A : if (!input_data)
 next_state =SEQ_B ;
 else
 next_state =SEQ_A ;

 SEQ_B : if (input_data)
 next_state = SEQ_C ;
 else
 next_state =IDLE ;

 SEQ_C : if (!input_data)
 next_state = SEQ_D;
 else
 next_state=SEQ_A ;
```

```
 SEQ_D : if (input_data )
 next_state = SEQ_A;
 else
 next_state = IDLE ;
 default : next_state = IDLE ;
 endcase
end

//------------CURRENT_STATE_LOGIC------------------------//
always@(posedge clk or negedge reset_n)
begin
 if (!reset_n) begin
 curr_state<=3'd0 ;
 end
else begin

 curr_state<=next_state;
 end
end

//-----------output_logic-------------------------//
assign seq_detected = (curr_state==SEQ_D && input_data);
endmodule
```

Test bench module fsm_tb

```
Inputs: Nil
Outputs: Nil
Function: The test bench applies random values and detects the
sequence. The waveform fsm_tb.vcd can be observed using wave-
form viewer.
Test bench file: fsm_tb.v

module fsm_tb;
reg Clk;
reg Reset_n;
reg [8:0] pattern;
reg data_in;
wire seq_detected;

//clock generation
 always #5 Clk = ~Clk;

 initial
```

```
begin
 Clk = 0;
 Reset_n = 1;

 $display("--------- Test Started ---------");
 #10 Reset_n = 0;
 #10 Reset_n = 1;

$display("--------- Sending Data pattern 111010101 ---------");
@ (posedge Clk);
 pattern = 9'b111010101;

 #10 data_in = pattern[8];
#10 data_in = pattern[7];
#10 data_in = pattern[6];
#10 data_in = pattern[5];
#10 data_in = pattern[4];
#10 data_in = pattern[3];
#10 data_in = pattern[2];
#10 data_in = pattern[1];
#10 data_in = pattern[0];

$display("--------- Sending Data pattern 110010101 ---------");
 @ (posedge Clk);
 pattern = 9'b110010101;
 data_in = pattern[8];
#10 data_in = pattern[7];
#10 data_in = pattern[6];
#10 data_in = pattern[5];
#10 data_in = pattern[4];
#10 data_in = pattern[3];
#10 data_in = pattern[2];
#10 data_in = pattern[1];
#10 data_in = pattern[0];

$display("--------- Sending Data pattern 101010101 ---------");

 pattern = 9'b101010101;
@ (posedge Clk);
#10 data_in = pattern[8];
#10 data_in = pattern[7];
#10 data_in = pattern[6];
#10 data_in = pattern[5];
#10 data_in = pattern[4];
```

```
#10 data_in = pattern[3];
#10 data_in = pattern[2];
#10 data_in = pattern[1];
#10 data_in = pattern[0];

$display("--------- Test Ended ---------");
 #1000 $finish;
end

fsm u_fsm(
 .clk(Clk), // Clock input of the design
 .reset_n(Reset_n),// active low, synchronous Reset input
 .input_data(data_in),// Input data bit.
 .seq_detected(seq_detected)// sequence detected
 );// End of port list

initial
begin
$dumpfile("fsm_tb.vcd");
$dumpvars(0,fsm_tb);
end
endmodule
```

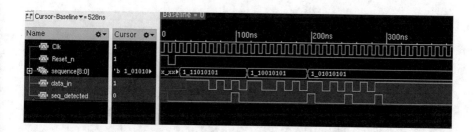

Linear feedback shift register

```
Polynomial 1+x+ x4
Inputs: en
Outputs: count_4
Function: The design works for polynomial 1+x+x4. The output gen-
erates pseudorandom numbers {count_4}. We cannot predict next count.
Design file: lfsr.v
/***************************************
// Module works for the polynomial 1+x+x4. //
// This is sequential block which require clock and reset //
//User can refer to any Verilog HDL language book to understand the
syntax of commands. //
***************************************/
```

```
module lfsr(
 clk,
 en,
 reset_n,
 count
 );
input clk;
input reset_n;
input en;
output [3:0] count;
 reg [3:0] count;
wire feedback;

 assign feedback =(count[3]^count[0]);
 always @(posedge clk or negedge reset_n)
 begin
 if(! reset_n)
 count =4'd1;
 else
 if(en)
 count ={count[2:0],feedback};
 end
endmodule
```

Test bench module lfsr_tb

Inputs: Nil
Outputs: Nil
*Function: The test bench applies random values and detect the 4-bit
counter output for polynomial 1+x+x4 . The waveform lfsr_tb.vcd
can be observed using waveform viewer.*

Test bench file: lfsr_tb.v

```
module lfsr_tb();
reg clk;
reg reset_n;
reg en;
wire [3:0] count;

lfsr u1 (
 .clk(clk),
 .reset_n(reset_n),
 .en(en),
```

```verilog
 .count(count)
 );
initial begin
clk=0;
forever #5 clk=~clk;
end

initial begin
#10;
@(posedge clk)
reset_n =0;
en=0;
#10;
 reset_n =1;
en=1;
#100 $finish;
end
initial begin
$dumpfile("lfsr_tb.vcd");
$dumpvars(0,lfsr_tb);
end
endmodule
```

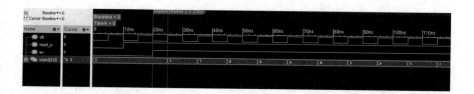

Hour-minute-second timer

```
Inputs: en
Outputs: sec_6,min_6,hour_5.
Function: When reset is high, all second, minute, and hour becomes
zeros. When reset is 0, second starts incrementing if second = 59
second becomes zero and minutes start incrementing, when minutes =
59 minutes become 0 and hours start incrementing.
Design file: timer.v
/***************************************
// Module increments second followed minutes followed by hours.
// This is sequential block which require clock and reset //
//User can refer to any Verilog HDL language book to understand the
syntax of commands. //
***************************************/
```

```verilog
module timer(
 clk, // input clock
 second,// second output
 minute,// minute output
 hour, // hour output
 rstn // active low reset
 );
//--------------------input ports------------------
 input clk;
 input rstn;
 //-------------------output ports----------------
 output [5:0] second;
 output [5:0] minute;
 output [4:0] hour;
//-----------------input datatype---------------
 wire clk;


//-----------------output datatype--------------


 reg [5:0] second;
 reg [5:0] minute;
 reg [4:0] hour;

 //this block starts for every posedge of the clock
 always @(posedge clk)
 begin
 if(rstn) // for every rising edge of the clock if reset is 1 load
 0 to second minute hour
 begin
 second <=6'd0;
 minute <= 6'd0;
 hour <= 5'd0;
 end

  else if (second == 6'd59)
  begin
  second <= 6'd0;// check if second = 59 reset second to zero
  if (minute == 6'd59)
  begin
  minute <= 6'd0;// check if minute = 59 reset minute to zero
  if (hour == 5'd23)
  begin
  hour <= 5'd0;//check if hour = 23 reset hour to zero
  end
```

```
    else
    begin
    hour <= hour + 5'd1;
    end
    end
    else
    begin
    minute <= minute + 6'd1;
    end
    end
    else
    begin
    second <= second + 6'd1;
    end
    end

endmodule
```

Test bench module timer_tb

Inputs: Nil
Outputs: Nil
Function: The test bench applies random values and checks the
results. The waveform timer_tb.vcd can be observed using wave-
form viewer.

Test bench file: timer_tb.v

```
module timer_tb;


 // Inputs
 reg clk;
 reg rstn;

 // Outputs
 wire [5:0] second;
 wire [5:0] minute;
 wire [4:0] hour;

 // clock generation
always #5 clk = ~clk; // toggle clock for every 5 ticks
```

```
initial begin
// Initialize Inputs
clk = 0;
rstn = 1;

//$display("--------- Test Started ---------");
#10 rstn = 1;
#10 rstn = 0;

#3000000 $finish;
end

timer uut (
.clk(clk),
.second(second),
.minute(minute),
.hour(hour),
.rstn(rstn)
);

initial
begin
$dumpfile("timer_tb.vcd");
$dumpvars(0,timer_tb);
end
endmodule
```

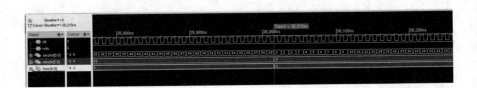

Self-sync scrambler

```
Inputs: bit_in
Outputs: bit_out
Function: This is a 7-bit scrambler for 802.11b synchronous active
high reset and with active high enable signal.
Design file: self_sync_scrambler.v,
/**************************************
// Module performs linear feedback shift register for 1+x3+x6
// This is sequential block which require clock and reset //
//User can refer to any Verilog HDL language book to understand the
syntax of commands. //
**************************************/
```

```verilog
module self_sync_scrambler (
  clock , // Clock input of the design
  resetn , // active low, synchronous Reset input
  enable , // Active high enable signal
  bit_in, // Input data bit.
  bit_out // Scrambled output bit.
  ); // End of port list
  //------------Input Ports-----------------------------
  input clock ;
  input resetn ;
  input enable ;
  input bit_in;

  //------------Output Ports----------------------------
  output bit_out;

  //------------Input ports Data Type-------------------
  // By rule all the input ports should be wires
  wire clock ;
  wire resetn ;
  wire enable ;
  //------------Output Ports Data Type------------------
  // Output port can be a storage element (reg) or a wire
  reg [6:0] state_out ;
  wire bit_out;

  //-----------Code Starts Here-------------------------
  assign feedback = (bit_in ^ state_out[6] ^ state_out[3]);
  assign bit_out = feedback;

  // We trigger the below block with respect to positive
  // edge of the clock.
  always @ (negedge resetn or posedge clock)
  begin : SCRAMBLER // Block Name
  if (resetn == 1'b0) begin
  state_out <= #1 7'b1111111;
  end
  // If enable is active, then we tick the state.
  else if (enable == 1'b1) begin
  state_out <= {state_out[5:0], feedback};
  end
  end // block: SCRAMBLER

endmodule
```

```
Design file: self_sync_descrambler.v
Inputs: bit_in
Outputs: bit_out
Function: This is a 7-bit descrambler for 802.11b synchronous
active high reset and with active high enable signal.
/***************************************
// Module performs linear feedback shift register for 1+x3+x6
// This is sequential block which require clock and reset //
//User can refer to any Verilog HDL language book to understand the
syntax of commands. //
***************************************/
module self_sync_descrambler (
 clock , // Clock input of the design
 resetn , // active high, synchronous Reset input
 enable , // Active high enable signal
 bit_in, // Input data bit.
 bit_out // Scrambled output bit.
 ); // End of port list
//-------------Input Ports----------------------------
input clock ;
input resetn ;
input enable ;
input bit_in;

//-------------Output Ports----------------------------
output bit_out;

//-------------Input ports Data Type------------------
// By rule all the input ports should be wires
wire clock ;
wire resetn ;
wire enable ;
//-------------Output Ports Data Type------------------
// Output port can be a storage element (reg) or a wire
reg [6:0] state_out ;
reg bit_out;

//-----------Code Starts Here------------------------
assign feedback = (bit_in ^ state_out[6] ^ state_out[3]);

// We trigger the below block with respect to positive
// edge of the clock.
always @ (negedge resetn or posedge clock)
begin : DESCRAMBLER // Block Name
if (resetn == 1'b0) begin
```

```
//Self synching, so a reset should be to the unknown state.
//This might cause a problem in synthesis.
state_out <= #1  7'bXXXXXXX;
end
else if (enable == 1'b1) begin
state_out <= {state_out[5:0],bit_in};
bit_out <= feedback;
end
end // block: DESCRAMBLER
endmodule
```

Test bench module self_sync_scr_tb_top

```
Inputs: Nil
Outputs: Nil
Function: The test bench applies random values for pattern and
checks the results by generating match signal. The waveform self_
sync_scr_tb_top.vcd can be observed using waveform viewer.

Test bench file: self_sync_scr_tb_top.v
Moduleself_sync_scr_tb_top;

reg Clk;
reg Resetn;
reg Enb;
reg [7:0] Pattern;
reg [7:0] DataIn;
reg [7:0] DataOut;
integer errCnt;
integer CompFlag;
reg Match;
wire Din;
wire Sout;
wire Dout;

//clock generation
 always #5 Clk = ~Clk;

assign Din = DataIn[7];


initial
begin
 Clk = 0;
```

```
Resetn = 1;
Enb = 0;
CompFlag =0;
errCnt = 0;
Match = 0;
$display("--------- Test Started ---------");
#10 Resetn = 0;
#10 Resetn = 1;

$display("--------- Sending Data Patternn : 0x55 ---------");
repeat (10) @ (posedge Clk);
Pattern = 8'h55;
DataIn = Pattern;
#10 Enb = 1;
repeat (100) begin
@ (posedge Clk) #1 DataIn = {DataIn[6:0],DataIn[7]};
end
repeat (10) @ (posedge Clk)Enb = 0;

$display("--------- Sending Data Patternn : 0x11 ---------");
repeat (10) @ (posedge Clk);
Enb = 1;
Pattern = 8'h11;
DataIn = Pattern;
repeat (100) begin
@ (posedge Clk) #1 DataIn = {DataIn[6:0],DataIn[7]};
end
repeat (10) @ (posedge Clk)Enb = 0;
CompFlag = 0;

$display("--------- Sending Data Patternn : 0x22 ---------");
repeat (10) @ (posedge Clk);
Enb = 1;
Pattern = 8'h22;
DataIn = Pattern;
repeat (100) begin
@ (posedge Clk) #1 DataIn = {DataIn[6:0],DataIn[7]};
end
repeat (10) @ (posedge Clk)Enb = 0;
CompFlag = 0;

$display("--------- Sending Data Patternn : 0x33 ---------");
repeat (10) @ (posedge Clk);
Enb = 1;
Pattern = 8'h33;
```

```
DataIn = Pattern;
repeat (100) begin
@ (posedge Clk) #1 DataIn = {DataIn[6:0],DataIn[7]};
end
repeat (10) @ (posedge Clk)Enb = 0;
CompFlag = 0;

$display("--------- Sending Data Patternn : 0x44 ---------");
repeat (10) @ (posedge Clk);
Enb = 1;
Pattern = 8'h44;
DataIn = Pattern;
repeat (100) begin
@ (posedge Clk) #1 DataIn = {DataIn[6:0],DataIn[7]};
end
repeat (10) @ (posedge Clk)Enb = 0;
CompFlag = 0;

$display("--------- Test Ended ---------");
#1000 $finish;
end

always@(posedge Clk)
begin
 if(Enb) begin
 DataOut = {DataOut[6:0],Dout};
 #1 if(DataOut == Pattern) Match = 1;
 else Match = 0;
 end
 else DataOut = 8'hXX;
end

self_sync_scrambler u_scarmb(
 .clock (Clk), // Clock input of the design
 .resetn (Resetn), // active low, synchronous Reset input
 .enable (Enb), // Active high enable signal
 .bit_in (Din) , // Input data bit.
 .bit_out (Sout) // Scrambled output bit.
 ); // End of port list

self_sync_descrambler u_descramb(
 .clock (Clk), // Clock input of the design
 .resetn (Resetn), // active low, synchronous Reset input
 .enable (Enb), // Active high enable signal
 .bit_in (Sout), // Input data bit.
```

```
.bit_out (Dout) // De-Scrambled output bit.
); // End of port list

initial
begin
$dumpfile("self_sync_scr_tb_top.vcd");
$dumpvars(0,self_sync_scr_tb_top);
end
endmodule
```

Side stream scrambler

```
Inputs: bit_in
Outputs: bit_out
Function: This is a 32-bit scrambler for 802.11b synchronous active
high reset and with active high enable signal.
Design file: side_stream_scrambler.v,
/****************************************
// Module performs lfsr for 1+x12+x32
// This is sequential block which require clock and reset //
//User can refer to any Verilog HDL language book to understand the
syntax of commands. //
****************************************/
module side_stream_scrambler ( clk ,
 reset_n ,
 en ,
 init_seed ,
 data_in ,
 data_out ,
 data_out_valid
 );

input clk ,
 reset_n ;

input en ;
```

```verilog
input [32:0] init_seed ;


input data_in ;


output reg data_out ,
 data_out_valid ;

reg [32:0] data_out_reg ;
wire xor_value1;
always@(posedge clk or negedge reset_n)
 begin
 if (!reset_n) begin
 data_out_reg<=33'd0;
 data_out_valid<=1'd0;
 end
 else begin
 data_out_valid<=en;
 data_out<=xor_value1;
 if (en)
 data_out_reg<={data_out_reg[31:0],xor_value1};
 else
 data_out_reg<=init_seed;
 end
end

assign xor_value= (data_out_reg[32]^data_out_reg[12]);


assign xor_value1=(data_in^xor_value);
endmodule

Design file: side_stream_descrambler.v
Inputs: bit_in
Outputs: bit_out
Function: This is a 32bit descrambler for 802.11b Synchronous
active high reset and with active high enable signal
/*************************************
// Module performs lfsr for 1+x12+x32
// This is sequential block which require clock and reset //
//User can refer to any Verilog HDL language book to understand the
syntax of commands. //
*************************************/
module side_stream_descrambler ( clk ,
```

```verilog
reset_n ,
en ,
init_seed ,
data_in ,
data_out ,
data_out_valid
);

input clk ,
reset_n ;

input en ;


input [32:0] init_seed ;


input data_in ;


output reg data_out ,
data_out_valid ;

reg [32:0] data_out_reg ;
wire xor_value1;
always@(posedge clk or negedge reset_n)
begin
if (!reset_n) begin
data_out_reg<=33'd0;
data_out_valid<=1'd0;
end
else begin
data_out_valid<=en;
data_out<=xor_value1;
if (en)
data_out_reg<={data_out_reg[31:0],data_in};
else
data_out_reg<=init_seed;
end
end

assign xor_value= (data_out_reg[32]^data_out_reg[12]);
assign xor_value1=(data_in^xor_value);

endmodule
```

Test bench module side_stream_scr_tb_top

Inputs: Nil
Outputs: Nil
Function: : The test bench applies random values for pattern and checks the results by generating match signal. The waveform tb_ top.vcd can be observed using waveform viewer.
Test bench file: side_stream_scr_tb.v

```
module side_stream_scr_tb_top;
reg Clk;
reg Resetn;
reg Enb;
reg [32:0] Pattern;
reg [32:0] DataIn;
reg [32:0] DataOut;
integer errCnt;
integer CompFlag;
reg Match;
wire Din;
wire Sout;
wire Dout;

//clock generation
 always #5 Clk = ~Clk;
 assign Din = DataIn[32];

initial
begin
 Clk = 0;
 Resetn = 1;
 Enb = 0;
 CompFlag =0;
 errCnt = 0;
 Match = 0;
 $display("--------- Test Started ---------");
 #10 Resetn = 0;
 #10 Resetn = 1;

 $display("--------- Sending Data Patternn : 0x55 ---------");
 repeat (1) @ (posedge Clk);
 Pattern = 33'h155555555;
 DataIn = Pattern;
 #1 Enb = 1;
 repeat (100) begin
```

```verilog
@ (posedge Clk) #5 DataIn = {DataIn[31:0],DataIn[32]};
end
//repeat (10) @ (posedge Clk)Enb = 0;

$display("--------- Sending Data Patternn : 0x11 ---------");
repeat (10) @ (posedge Clk);
Enb = 1;
Pattern = 33'h111111111;
DataIn = Pattern;
repeat (100) begin
@ (posedge Clk) #5 DataIn = {DataIn[31:0],DataIn[32]};
end
//repeat (10) @ (posedge Clk)Enb = 0;
CompFlag = 0;

$display("--------- Sending Data Patternn : 0x22 ---------");
repeat (10) @ (posedge Clk);
Enb = 1;
Pattern = 33'h122222222;
DataIn = Pattern;
repeat (100) begin
@ (posedge Clk) #5 DataIn = {DataIn[31:0],DataIn[32]};
end

CompFlag = 0;


$display("--------- Sending Data Patternn : 0x33 ---------");
repeat (10) @ (posedge Clk);
Enb = 1;
Pattern = 33'h133333333;
DataIn = Pattern;
repeat (100) begin
@ (posedge Clk) #1 DataIn = {DataIn[31:0],DataIn[32]};
end
// repeat (10) @ (posedge Clk)Enb = 0;
CompFlag = 0;

$display("--------- Sending Data Patternn : 0x44 ---------");
repeat (10) @ (posedge Clk);
Enb = 1;
Pattern = 33'h144444444;
DataIn = Pattern;
repeat (100) begin
```

```verilog
 @ (posedge Clk) #1 DataIn = {DataIn[31:0],DataIn[32]};
 end
 CompFlag = 0;

 $display("--------- Test Ended ---------");
 #10000 $finish;
end


always@(posedge Clk)
begin
 if(Enb) begin
 DataOut = {DataOut[32:0],Dout};
 #1 if(DataOut == Pattern) Match = 1;
 else Match = 0;
 end
 else DataOut = 33'hXXXXXXXX;
end

 side_stream_scrambler u1( .clk (Clk) ,
 .reset_n(Resetn) ,
 .en (Enb) ,
 .init_seed (33'h155555555) ,
 .data_in (Din) ,
 .data_out (Sout) ,
 .data_out_valid ()
 );

 side_stream_descrambler u2( .clk (Clk) ,
 .reset_n(Resetn) ,
 .en (Enb) ,
 .init_seed (33'hXXXXXXXX) ,
 .data_in (Sout) ,
 .data_out (Dout) ,
 .data_out_valid ()
 );

 initial
 begin
 $dumpfile("side_stream_scr_tb_top.vcd");
 $dumpvars(0,side_stream_scr_tb_top);
 end
endmodule
```

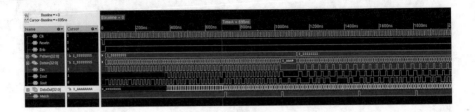

Colored ball puzzle box

Inputs: red_blue_vld
Outputs: number of chance valid, number of chance count, pickup of
ball is wrong, pickup ball from red blue box
Function: This works based on FSM if current state being idle and
config interface being high and then ball pickup from red-blue box
will be high. If current state being OUT_put state, then number of
chance valid will be high. If current state being error_state,
output pickup ball is wrong is high.
Design file: puzzle.v,
/***************************************
// Module works based on FSM
// This is sequential block which require clock and reset //
//User can refer to any Verilog HDL language book to understand the
syntax of commands. //
***************************************/
module puzzle_3box (
//-----------------global_interface-----------------------//
  clk ,
  reset_n ,
  cfg_start_algo , //config interface
//----------------Input_interface-----------------------//
  red_blue_vld ,
//-----------------output_interface----------------------//
  ball_pickup_from_red_blue_box ,
  number_of_chance_vld ,
  number_of_chance_count ,
  pickup_of_ball_is_wrong
  );
//-----------------global_interface----------------------//
input clk ,
  reset_n ;
input cfg_start_algo ;
//----------------Input_interface----------------------//
input red_blue_vld ;
//-----------------output_interface----------------------//
output number_of_chance_vld ;

```
output reg [31:0] number_of_chance_count ;
output ball_pickup_from_red_blue_box ,
 pickup_of_ball_is_wrong ;

reg [1:0] curr_state, next_state;


parameter IDLE = 2'd0 ,
 PICKUP_RED_BLUE = 2'd1 ,
 OUTPUT_STATE = 2'd2 ,
 ERROR_STATE = 2'd3 ;

//-------------------next_state_logic-------------------
------//


always@( cfg_start_algo ,
 red_blue_vld
 )
 begin
 case (curr_state)

 IDLE : if (cfg_start_algo)
 next_state= PICKUP_RED_BLUE;
 else
 next_state = IDLE ;

 PICKUP_RED_BLUE : if ( red_blue_vld )
 next_state = OUTPUT_STATE ;
 else
 next_state = ERROR_STATE;
 OUTPUT_STATE : next_state= IDLE ;
 ERROR_STATE : next_state = IDLE;
 default : next_state =IDLE ;
 endcase
end

always@(posedge clk or negedge reset_n)
 begin
 if (!reset_n) begin
 curr_state=2'd0 ;
 number_of_chance_count<=32'd0;
 end
 else begin
 curr_state<=next_state ;
```

```
 if (curr_state== PICKUP_RED_BLUE )
 number_of_chance_count<=number_of_chance_count+32'd1;
 else if (curr_state== OUTPUT_STATE)
 number_of_chance_count<=32'd0 ;


 end
end
assign  ball_pickup_from_red_blue_box  =  (curr_state  ==  IDLE  &&
cfg_start_algo);
assign number_of_chance_vld = (curr_state==OUTPUT_STATE) ;
assign pickup_of_ball_is_wrong = (curr_state ==ERROR_STATE) ;
endmodule
```

Test bench module puzzle3box_tb

```
Inputs: Nil
Outputs: Nil
Function: The test bench applies random values of input and checks
for the result. The waveform puzzle3box_tb.vcd can be observed
using waveform viewer.
Test bench file: puzzle3box_tb.v
module puzzle3box_tb;
reg clk;
reg reset_n;
reg cfg_start_algo;
reg red_blue_vld;

wire [31:0] number_of_chance_count;
wire number_of_chance_vld;
wire pickup_of_ball_is_wrong;
wire ball_pickup_from_red_blue_box;

always #5 clk=~clk;
initial begin
clk =0;
reset_n = 0;
cfg_start_algo = 0;
red_blue_vld = 0;

#10 reset_n =0;
#10 reset_n =1;
  cfg_start_algo = 1;
```

```
#10 red_blue_vld =1;


#10 cfg_start_algo = 0;
#10 cfg_start_algo = 1;
#10 red_blue_vld =0;

#100 $finish;
end
puzzle_3box uut (
 .clk(clk),
 .reset_n(reset_n),
 .cfg_start_algo(cfg_start_algo),
 .red_blue_vld(red_blue_vld),
 .ball_pickup_from_red_blue_box(ball_pickup_from_red_blue_box),
 .number_of_chance_vld(number_of_chance_vld),
 .number_of_chance_count(number_of_chance_count),
 .pickup_of_ball_is_wrong(pickup_of_ball_is_wrong)
 );

initial begin
$dumpfile("puzzle3box_tb.vcd");
$dumpvars(0,puzzle3box_tb);
end
endmodule
```

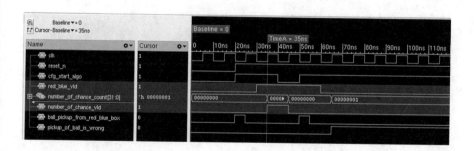

Scratchpad registers

```
Inputs: addr_sel, wr_rd_addr_3, write_data_32
Outputs: read_data_32
Function: 8-location 32-bit scratchpad resister set. The design
reads the data written at the particular address.
Design file: scratch_pad_reg.v
/**************************************
```

```verilog
// Module reads the 32_data written from the 3-bit address.
// This is sequential block which require clock and reset //
//User can refer to any Verilog HDL language book to understand the
syntax of. //
// commands
**************************************/
module scratch_pad_reg(
//-----------------clock_reset----------------//
 clk ,
 reset_n ,
//---------------SW_INTERFACE--------------------//
 addr_sel ,
 wr_rd_addr ,
 write_en ,
 read_en ,
 write_data ,
 read_data
 );

//-----------------clock_reset----------------//
input clk ,
 reset_n ;
//---------------SW_INTERFACE-------------------//
input addr_sel ;

input [2:0] wr_rd_addr ;


input write_en ,
 read_en ;

input [31:0] write_data ;


output [31:0] read_data ;


reg [31:0] reg0 ,
 reg1 ,
 reg2 ,
 reg3 ,
 reg4 ,
 reg5 ,
 reg6 ;
```

```verilog
wire sel0 ,
 sel1 ,
 sel2 ,
 sel3 ,
 sel4 ,
 sel5 ,
 sel6 ;

assign sel0 = (addr_sel && wr_rd_addr==3'd0) ;
assign sel1 = (addr_sel && wr_rd_addr==3'd1) ;
assign sel2 = (addr_sel && wr_rd_addr==3'd2) ;
assign sel3 = (addr_sel && wr_rd_addr==3'd3) ;
assign sel4 = (addr_sel && wr_rd_addr==3'd4) ;
assign sel5 = (addr_sel && wr_rd_addr==3'd5) ;
assign sel6 = (addr_sel && wr_rd_addr==3'd6) ;

assign read_data = (sel0 && read_en) ? reg0 :
 (sel1 && read_en) ? reg1 :
 (sel2 && read_en) ? reg2 :
 (sel3 && read_en) ? reg3 :
 (sel4 && read_en) ? reg4 :
 (sel5 && read_en) ? reg5 : reg6 ;

always@(posedge clk or negedge reset_n)
begin
 if (!read_en) begin
 reg0<=32'd0;
 end
 else begin

 if (write_en && sel0)


 reg0<=write_data ;
 end
end

always@(posedge clk or negedge reset_n)
begin
 if (!read_en) begin
 reg1<=32'd0;
 end
 else begin

 if (write_en && sel1)
```

```verilog
 reg1<=write_data ;
 end
end

always@(posedge clk or negedge reset_n)
begin
 if (!read_en) begin
 reg2<=32'd0;
 end
 else begin

 if (write_en && sel2)


 reg2<=write_data ;
 end
end

always@(posedge clk or negedge reset_n)
begin
 if (!read_en) begin
 reg3<=32'd0;
 end
 else begin

 if (write_en && sel3)


 reg3<=write_data ;
 end
end

always@(posedge clk or negedge reset_n)
begin
 if (!read_en) begin
 reg4<=32'd0;
 end
 else begin

 if (write_en && sel4)


 reg4<=write_data ;
 end
end
```

```verilog
always@(posedge clk or negedge reset_n)
begin
 if (!read_en) begin
 reg5<=32'd0;
 end
 else begin

 if (write_en && sel5)


 reg5<=write_data ;
 end
end

always@(posedge clk or negedge reset_n)
begin
 if (!read_en) begin
 reg6<=32'd0;
 end
 else begin

 if (write_en && sel6)


 reg6<=write_data ;
 end
end

endmodule
```

Test bench module scratch_pad_reg_tb

```
Inputs: Nil
Outputs: Nil
Function: The test bench applies random values of input and checks
for the result. The waveform scratch_pad_reg_tb.vcd can be observed
using waveform viewer.

Test bench file: scratch_pad_reg_tb.v


module scratch_pad_reg_tb;
reg clk;
reg reset_n ;
```

```
reg en;
reg addr_sel;
reg [2:0] wr_rd_addr ;
reg write_en;
reg read_en;
reg [31:0] write_data;

wire [31:0] read_data;


always #5 clk=~clk;


initial
 begin
 clk=0;
 reset_n = 0;
 en = 0;

 #10 reset_n = 0;
 #10 reset_n = 1;
 en=1;

 addr_sel=1; wr_rd_addr=000; write_en=1; read_en=1;
#10 addr_sel=1; wr_rd_addr=001; write_en=1; write_data=32'h11111111;
read_en=1;
#10 addr_sel=1; wr_rd_addr=010; write_en=1; write_data=32'h22222222;
read_en=1;
#10 addr_sel=1; wr_rd_addr=011; write_en=1; write_data=32'h33333333;
read_en=1;
#10 addr_sel=1; wr_rd_addr=100; write_en=1; write_data=32'h44444444;
read_en=1;
#10 addr_sel=1; wr_rd_addr=101; write_en=1; write_data=32'h55555555;
read_en=1;
#10 addr_sel=1; wr_rd_addr=110; write_en=1; write_data=32'h66666666;
read_en=1;
#10 addr_sel=0; wr_rd_addr=000; write_en=1; write_data=32'h77777777;
read_en=1;
#10 addr_sel=1; wr_rd_addr=110; write_en=1; write_data=32'h88888888;
read_en=1;
#10 addr_sel=0; wr_rd_addr=110; write_en=1; write_data=32'h99999999;
read_en=1;
#100 $finish;
end
```

```
scratch_pad_reg uut (
 .clk(clk),
 .reset_n(reset_n),
 .addr_sel(addr_sel),
 .wr_rd_addr(wr_rd_addr),
 .write_en(write_en),
 .read_en(read_en),
 .write_data(write_data),
 .read_data(read_data)
 );

initial
 begin
 $dumpfile("scratch_pad_reg_tb.vcd");
 $dumpvars(0,scratch_pad_reg_tb);
 end
endmodule
```

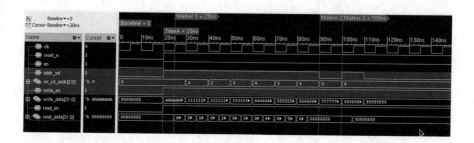

Configuration register

Inputs: addr_sel, wr_rd_addr_3, write_data_32
Outputs: read_data_32, reg0_32,reg1_32,reg2_32,reg3_32,reg4_32,r
eg5_32,reg6_32
Function: The design reads the data written at the particular
address and also stores the data in 32-bit register for respective
address.
Design file: config_reg.v,
/***************************************
// Module reads the 32_data written from the 3-bit address. And
stores the data in 32bit register
// This is sequential block which require clock and reset //
//User can refer to any Verilog HDL language book to understand the
syntax of commands. //
***************************************/
module config_reg (

```verilog
//------------------clock_reset-----------------//
 clk ,
 reset_n ,
//---------------SW_INTERFACE--------------------//
 addr_sel ,
 wr_rd_addr ,
 write_en ,
 read_en ,
 write_data ,
 read_data,
//----------------OUTPUT------------------------//
 reg0 ,
 reg1 ,
 reg2 ,
 reg3 ,
 reg4 ,
 reg5 ,
 reg6

 );


//------------------clock_reset-----------------//
input clk ,
 reset_n ;
//---------------SW_INTERFACE--------------------//
input addr_sel ;

input [2:0] wr_rd_addr ;


input write_en ,
 read_en ;

input [31:0] write_data ;
output [31:0] read_data ;

output reg [31:0] reg0 ,
 reg1 ,
 reg2 ,
 reg3 ,
 reg4 ,
 reg5 ,
 reg6 ;
```

```verilog
wire sel0 ,
 sel1 ,
 sel2 ,
 sel3 ,
 sel4 ,
 sel5 ,
 sel6 ;


assign sel0 = (addr_sel && wr_rd_addr==3'd0) ;
assign sel1 = (addr_sel && wr_rd_addr==3'd1) ;
assign sel2 = (addr_sel && wr_rd_addr==3'd2) ;
assign sel3 = (addr_sel && wr_rd_addr==3'd3) ;
assign sel4 = (addr_sel && wr_rd_addr==3'd4) ;
assign sel5 = (addr_sel && wr_rd_addr==3'd5) ;
assign sel6 = (addr_sel && wr_rd_addr==3'd6) ;

assign read_data = (sel0 && read_en) ? reg0 :
 (sel1 && read_en) ? reg1 :
 (sel2 && read_en) ? reg2 :
 (sel3 && read_en) ? reg3 :
 (sel4 && read_en) ? reg4 :
 (sel5 && read_en) ? reg5 : reg6 ;

always@(posedge clk or negedge reset_n)
begin
 if (!read_en) begin
 reg0<=32'd0;
 end
 else begin

 if (write_en && sel0)


 reg0<=write_data ;
 end
end

always@(posedge clk or negedge reset_n)
begin
 if (!read_en) begin
 reg1<=32'd0;
 end
 else begin

 if (write_en && sel1)
```

```verilog
 reg1<=write_data ;
 end
end

always@(posedge clk or negedge reset_n)
begin
 if (!read_en) begin
 reg2<=32'd0;
 end
 else begin
 if (write_en && sel2)

 reg2<=write_data ;
 end
end

always@(posedge clk or negedge reset_n)
begin
 if (!read_en) begin
 reg3<=32'd0;
 end
 else begin

 if (write_en && sel3)

 reg3<=write_data ;
 end
end

always@(posedge clk or negedge reset_n)
begin
 if (!read_en) begin
 reg4<=32'd0;
 end
 else begin

 if (write_en && sel4)

 reg4<=write_data ;
 end
end

 always@(posedge clk or negedge reset_n)
```

```
begin
 if (!read_en) begin
 reg5<=32'd0;
 end
 else begin

 if (write_en && sel5)


 reg5<=write_data ;
 end
end

always@(posedge clk or negedge reset_n)
begin
 if (!read_en) begin
 reg6<=32'd0;
 end
 else begin

 if (write_en && sel6)


 reg6<=write_data ;
 end
end

endmodule
```

Test bench module config_reg_tb

Inputs: Nil
Outputs: Nil
Function: The test bench applies random values of input and checks for the result. The waveform config_reg_tb.vcd can be observed using waveform viewer.

Test bench file: config_reg_tb.v

```
module config_reg_tb();
reg clk;
reg reset_n;
reg addr_sel;
```

```verilog
reg [2:0]wr_rd_addr;
reg write_en ;
reg read_en ;
reg [31:0] write_data ;

wire [31:0] read_data;
wire [31:0] reg0;
wire[31:0] reg1;
wire [31:0]reg2;
wire [31:0]reg3;
wire [31:0]reg4;
wire[31:0] reg5;
wire[31:0] reg6;

initial begin
clk =0;
forever #5 clk =~clk;
end

config_reg u1 (
 .clk(clk),
 .reset_n(reset_n),
 .addr_sel(addr_sel),
 .wr_rd_addr(wr_rd_addr),
 .write_en(write_en),
 .read_en(read_en),
 .write_data(write_data),
 .read_data(read_data),
 .reg0(reg0),
 .reg1(reg1),
 .reg2(reg2),
 .reg3(reg3),
 .reg4(reg4),
 .reg5(reg5),
 .reg6(reg6));

 initial begin
 reset_n =0;
 addr_sel=0;
 wr_rd_addr=0;
 write_en=0;
 read_en=0;
 write_data=0;
```

```
#10 reset_n =1;


#10 addr_sel=1; wr_rd_addr=000; write_en=1; write_data=32'hAAAAAAAA;
read_en=1;
#10 addr_sel=1; wr_rd_addr=001; write_en=1; write_data=32'h11111111;
read_en=1;
#10 addr_sel=1; wr_rd_addr=010; write_en=1; write_data=32'h22222222;
read_en=1;
#10 addr_sel=1; wr_rd_addr=011; write_en=1; write_data=32'h33333333;
read_en=1;
#10 addr_sel=1; wr_rd_addr=100; write_en=1; write_data=32'h44444444;
read_en=1;
#10 addr_sel=1; wr_rd_addr=101; write_en=1; write_data=32'h55555555;
read_en=1;
#10 addr_sel=1; wr_rd_addr=110; write_en=1; write_data=32'h66666666;
read_en=1;

#100 $finish;
end
initial
 begin
 $dumpfile("config_reg_tb.vcd");
 $dumpvars(0,config_reg_tb);
 end

endmodule
```

Clock domain crossover

```
/*****************************************************
//// Description ////
//// Signals transfer from one clock to another clock domain ////
//// 1. Clocks can be asynchrnous or synchronous ////
//// 2. Clocks frequency may be smaller or greater ////
//// 3. Strobe signal out is always single cycle ////
//// 4. Up to 4 field signals can be synchronized ////
//// ////
//// ////
*****************************************************/

module clock_transfer #(
 parameter FIELD_WIDTH1 = 1,
 parameter FIELD_WIDTH2 = 1,
 parameter FIELD_WIDTH3 = 1,
 parameter FIELD_WIDTH4 = 1
 ) (
 reset_n,
 clk_in,
 strobe_in,
 field_in_1,
 field_in_2,
 field_in_3,
 field_in_4,

 clk_out,
 strobe_out,
 field_out_1,
 field_out_2,
 field_out_3,
 field_out_4
 );

input reset_n;
input clk_in;
input strobe_in;
input [FIELD_WIDTH1 - 1 : 0] field_in_1;
input [FIELD_WIDTH2 - 1 : 0] field_in_2;
input [FIELD_WIDTH3 - 1 : 0] field_in_3;
input [FIELD_WIDTH4 - 1 : 0] field_in_4;

input clk_out;
output strobe_out;
```

```verilog
output [FIELD_WIDTH1 - 1 : 0] field_out_1;
output [FIELD_WIDTH2 - 1 : 0] field_out_2;
output [FIELD_WIDTH3 - 1 : 0] field_out_3;
output [FIELD_WIDTH4 - 1 : 0] field_out_4;

reg strobe_in_d;
wire strobe_in_edge;
reg strobe_in_latch;
reg [FIELD_WIDTH1 - 1 : 0] field_latch_1;
reg [FIELD_WIDTH2 - 1 : 0] field_latch_2;
reg [FIELD_WIDTH3 - 1 : 0] field_latch_3;
reg [FIELD_WIDTH4 - 1 : 0] field_latch_4;
reg strobe_transfer_1;
reg strobe_transfer_2;
reg strobe_out;
reg [FIELD_WIDTH1 - 1 : 0] field_out_1;
reg [FIELD_WIDTH2 - 1 : 0] field_out_2;
reg [FIELD_WIDTH3 - 1 : 0] field_out_3;
reg [FIELD_WIDTH4 - 1 : 0] field_out_4;

//clk_out clocked FFs
reg strobe_reclocked_1;
reg strobe_reclocked_2;
reg strobe_reclocked_3;

//Delay strobe_in to allow edge detect
always @(posedge clk_in or negedge reset_n)
begin : del_p
 if (reset_n == 1'b0) strobe_in_d <= 1'b0;
 else strobe_in_d <= strobe_in;
end

// Edge detect to latch strobe itself and fields on rising edge.
assign strobe_in_edge = strobe_in & (~strobe_in_d);

//strobe_in_latch latches the incoming strobe, and is not cleared
until the
//logic has passed over the the outgoing clock domain.
always @(posedge clk_in or negedge reset_n)
begin : latch_in
 if (reset_n == 1'b0) begin
 strobe_in_latch <= 1'b0;
 strobe_transfer_1 <= 1'b0;
 strobe_transfer_2 <= 1'b0;
 end
```

```verilog
 else begin
  if (strobe_in_edge  ==  1'b1 && (strobe_transfer_1  ==  1'b1 ||
strobe_transfer_2 == 1'b1)) begin
  // $display ("Error: strobes are too close. Logic does not
function.\n");
  // $finish;
  end
  strobe_transfer_1 <= strobe_reclocked_2;
  strobe_transfer_2 <= strobe_transfer_1;
    strobe_in_latch   <=   strobe_in_edge  |   (strobe_in_latch  &
!(strobe_transfer_2));
  end
end

//Latch the field values on the incoming strobe
always @(posedge clk_in or negedge reset_n)
begin : latch_field
  if (reset_n == 1'b0) begin
  field_latch_1 <= 'b0;
  field_latch_2 <= 'b0;
  field_latch_3 <= 'b0;
  field_latch_4 <= 'b0;
  end
  else begin
  if (strobe_in_edge == 1'b1) begin
  field_latch_1 <= field_in_1;
  field_latch_2 <= field_in_2;
  field_latch_3 <= field_in_3;
  field_latch_4 <= field_in_4;
  end
  end
end

//Retime the signals into the outgoing clock domain and generate
the output signals.
//Note that field_out may partially or wholly change on the cycle
before strobe_out, but
//must only be inspected by the calling code on assertion of
strobe_out :

always @(posedge clk_out or negedge reset_n)
begin : gen_outputs
  if (reset_n == 1'b0) begin
  strobe_reclocked_1 <= 1'b0;
  strobe_reclocked_2 <= 1'b0;
```

```
  strobe_reclocked_3 <= 1'b0;
  strobe_out <= 1'b0;
  field_out_1 <= 'b0;
  field_out_2 <= 'b0;
  field_out_3 <= 'b0;
  field_out_4 <= 'b0;
  end
  else begin
  strobe_reclocked_1 <= strobe_in_latch; // Clock domain crossing.
  strobe_reclocked_2 <= strobe_reclocked_1;
  strobe_reclocked_3 <= strobe_reclocked_2;

  strobe_out <= strobe_reclocked_2 & !(strobe_reclocked_3);


  field_out_1 <= field_latch_1; // Clock domain crossing.
  field_out_2 <= field_latch_2;
  field_out_3 <= field_latch_3;
  field_out_4 <= field_latch_4;
  end
end
endmodule
```

Test bench module clock_transfer_tb_top

```
Inputs: Nil
Outputs: Nil
Function: The test bench applies random values of input fields and
sets strobe_in in clk_in and expects the fields to be transferred
to clk_out domain. The waveform clock_transfer.vcd can be observed
using waveform viewer.

Test bench file: clock_transfer_tb_top.v


module clock_transfer_tb_top;
reg reset_n,
 reg clk_in,
 reg strobe_in,
 reg field_in_1,
 reg field_in_2,
 reg field_in_3,
 reg field_in_4,
 reg clk_out,
```

```verilog
wire strobe_out;
wire field_out_1;
wire field_out_2;
wire field_out_3;
wire field_out_4;

//clock generation
 always #5 clk_in = ~clk_in;
 always #10 clk_out = ~clk_out;

initial
begin
 clk_in = 0;
 clk_out =0;
 reset_n= 1;
 strobe_in = 0;
 $display("--------- Test Started ---------");
 #10 reset_n = 0;
 #10 reset_n = 1;

 repeat (1) @ (posedge clk_in);
 field_in_1 = 1'b0;
 #1 field_in_2 = 1'b0;
 #1 field_in_3 = 1'b0;
 #1 field_in_4 = 1'b0;

 repeat (100) begin
 @ (posedge clk_in) #5 field_in_1 = 1'b1;
 strobe_in = 1'b1;
 @ (posedge clk_in) #5 field_in_2 = 1'b1;
 @ (posedge clk_in) #5 field_in_3 = 1'b1;
 @ (posedge clk_in) #5 field_in_4 = 1'b1;

 end


 clock_transfer
 uu1(
 .reset_n(reset_n),
 .clk_in(clk_in),
 .strobe_in(strobe_in),
 .field_in_1(field_in_1),
 .field_in_2(field_in_2),
```

```
.field_in_3(field_in_3,
.field_in_4(field_in_4,

.clk_out(clk_out),
.strobe_out(strobe_out),
.field_out_1(field_out_1),
.field_out_2(field_out_2),
.field_out_3(field_out_3),
.field_out_4(field_out_4)
);

initial
begin
$dumpfile("clock_transfer.vcd");
$dumpvars(2,clock_transfer_tb_top);
#1000 $finish;
end
endmodule
```

11.7 Part II

11.7.1 Design Flow

This section intends to take an example design and set up the synthesis and LEC flow. It contains design for simulation, synthesis, a constraint file used for synthesis, a synthesis script, dummy library file, and a logic equivalence check (LEC) script for RTL vs. gate. Other procedures in physical design require an EDA P&R tool environment where the design files and corresponding constraint files have to be imported and processed. Hence, the design flow with synthesis, simulation (given in Part 1), and LEC will set the minimum design flow to continue the design further. Advancement in the design flow actually requires library files with all EDA views of the standard cells and modules. A design example generating output shown in Fig. 11.2 is used to define the design flow. RTL modules with .v extension and the design constraint file .sdc are used as design inputs for the synthesis process, and netlist file with .vg extension is generated. The dummy library file in Liberty format (extract with .lib extension) and layout exchange format file (.lef file format) are given in this section for reference only and to demonstrate the flow. The user has to get access to actual library files for doing actual Synthesis, LEC, STA, and P&R. Executable scripts for synthesis and LEC are given for the design example. It is to be noted that the scripts can be customized to run on any design with suitable modifications and by replacing the correct commands from the chosen tools.

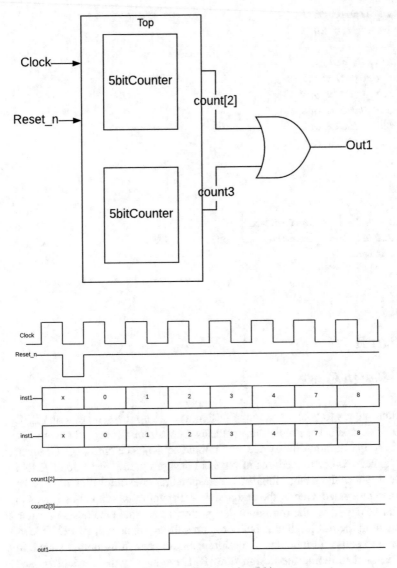

Fig. 11.2 Design example with timing diagram using 5-bit counters

Design file

```
################################################################
###########
This is the RTL module of a 5 bit counter design.This design will
be used to set the design flow.
The design modelled as RTL file
```

```
############################################################
###########

module 5bitcounter (clk, resetn, count);
 input clk, resetn;
 output [4:0] count;
 reg [4:0] count;

 always @(posedge clk or posedge resetn)
 begin
 if (~resetn)
 count <= 5'b00000;
 else
 count <= count + 1;
 end
endmodule

module top (clk, resetn, pso1, pso2, iso1_n, iso2_n, restore1,
restore2, out1);
 input clk, reset, pso1, pso2, iso1_n, iso2_n, restore1, restore2;
 output out1;
 wire [4:0] count1, count2;

 5bitcounter inst1(.clk(clk), .resetn(resetn), .count(count1));
 5bitcounter inst2(.clk(clk), .resetn(resetn), .count(count2));

 assign out1 = count1[3] || count2[2];


endmodule


Test bench for the example design:
module mycounter_t ;
wire [3:0] count;
reg resetn,clk;

initial
clk = 1'b0;

always
 #5 clk = ~clk;

top m1 ( (.clk(clk), resetn(resetn), out1(out1));
```

```
initial
begin
 resetn = 1'b1 ;

 #15 resetn =1'b0;
 #30 resetn =1'b1;
 #300 $finish;
end

initial
begin
 $dumpfile ("top.vcd");
 $dumpvars(2, top);
end

endmodule
```

##
#########

Design constraint file in standard delay constraint (SDC) format:

It is also called Synopsys design constraint file as it was defined by Synopsys. This file is a tool command language (TCL)-based script file and hence follows TCL command syntax. SDC contains mainly the following constraints that are very essential for design:

- Clock definition
- Generated clock (derived clock)
- Input-output delay
- Min/max delay
- False path
- Multicycle path
- Case analysis
- Disable timing arcs

For the design example, please refer to the timing needs shown in Fig. 11.3. Since it is pre-layout, the wire load model used is zero wire load where interconnect delays are not considered.

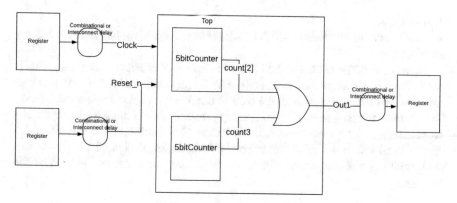

Fig. 11.3 Use case depicting design example with possible IO delays for definition in SDC

SDC file sample top.sdc is given below:

##
##########

```
set sdc_version 1.0
# define design top instance and units for parameters time and
capacitance
current_design top
set_units -time 1.0ns
set_units -capacitance 1000.0fF

# generation of clock
set_clock_gating_check -setup 0.0
create_clock -name "clk" -add -period 8.0 -waveform {0.0 4.0}
[get_ports clk]
# input-output delays expected for the design example
set_input_delay -clock [get_clocks clk] -add_delay 0.3 [get_
ports clk]
set input_delay 0.5 [get_ports resetn]
set_output_delay 0.8 [get_ports out1]

#set_ideal_net [get_nets iso1_n]
#set_ideal_network [get_ports iso1_n]

# pre-layout uses zero wire-load model
#set_wire_load_model "zero_wireload"
```

Library files
###
##########

Liberty files: The extract of the library file for a Adder cell is shown here. This is the dummy file to show the content of the lib file. It is required to have the fabricatable library of this type with all the cells to execute a process of synthesis. Liberty file contains for each logic cell, area, timing models, power models, timing checks to be used for the particular path in the circuit. The look up table contains three dimensional values of timing and internal power. In SoC design which uses library with multiple voltages, there will be corresponding liberty files for each of the voltage.

###
##########

```
.lib extract:
/* ------------------------- *
 * Design : ADDFHX2 *
 * ------------------------- */
cell (ADDFHX2) {
area : 8.208000;
cell_leakage_power : 0.327774;
rail_connection( VDD, RAIL_VDD );
rail_connection( VSS, RAIL_VSS );
pin(A) {
direction : input;
input_signal_level : RAIL_VDD;
capacitance : 0.00289594;
rise_capacitance : 0.00288999;
fall_capacitance : 0.00289594;
}
pin(B) {
# Data similar to pin(A)
}
pin(CI) {
# Data similar to pin(A)
}
pin(CO) {
direction : output;
output_signal_level : RAIL_VDD;
capacitance : 0;
rise_capacitance : 0;
fall_capacitance : 0;
max_capacitance : 0.262575;
function : "(((A B)+(B CI))+(CI A))";
timing() {
```

```
related_pin : "A";
timing_sense : positive_unate;
cell_rise(delay_template_3x3) {
index_1 ("0.008, 0.04, 0.08");
index_2 ("0.01, 0.06, 0.1");
values ( \
"0.205832, 0.395553, 0.539816", \
"0.217523, 0.407235, 0.55108 ", \
"0.232146, 0.421821, 0.565704 ");
}
rise_transition(delay_template_3x3) {
index_1 ("0.008, 0.04, 0.08");
index_2 ("0.01, 0.06, 0.1");
values ( \
"0.114013, 0.463975, 0.756059", \
"0.114164, 0.463936, 0.752876", \
"0.114441, 0.463654, 0.753174");
}
cell_fall(delay_template_3x3) {
index_1 ("0.008, 0.04, 0.08");
index_2 ("0.01, 0.06, 0.1");
values ( \
"0.199984, 0.415461, 0.580846", \
"0.211593, 0.42712, 0.592588", \
"0.225795, 0.441286, 0.606689");
}
fall_transition(delay_template_3x3) {
index_1 ("0.008, 0.04, 0.08");
index_2 ("0.01, 0.06, 0.1");
values ( \
"0.121746, 0.516895, 0.840346", \
"0.120985, 0.516002, 0.840337", \
"0.121692, 0.516881, 0.841414") ;
}
}
timing() {
related_pin : "B";
# Data similar to pin (A)
}
timing() {
related_pin : "CI";
# Data similar to pin (A)
}
internal_power() {
related_pin : "A";
```

```
rise_power(energy_template_3x3) {
index_1 ("0.008, 0.04, 0.08");
index_2 ("0.01, 0.06, 0.1");
values ( \
"0.002446, 0.002507, 0.002516", \
"0.002431, 0.002493, 0.002502", \
"0.002424, 0.002486, 0.002495");
}
fall_power(energy_template_3x3) {
index_1 ("0.008, 0.04, 0.08");
index_2 ("0.01, 0.06, 0.1");
values ( \
"0.002446, 0.002507, 0.002516", \
"0.002431, 0.002493, 0.002502", \
"0.002424, 0.002486, 0.002495");
}
internal_power() {
related_pin : "B";
# Data similar to pin(A)
}
}
}
pin(S) {
direction : output;
output_signal_level : RAIL_VDD;
capacitance : 0;
rise_capacitance : 0;
fall_capacitance : 0;
max_capacitance : 0.255238;
function : "((A^B)^CI)";
timing () {
# Timing Data similar to Pin (CO) with respect to related pins
A, B, CI
}
Internal_power() {
# Internal Power Data similar to Pin (CO) with respect to related
pins A, B, CI
}
}
}
```

Executable Scripts
```
############################################################
##########
```
Synthesis Tcl script:

Synthesis is tool dependent and hence the command syntax, can be different for different synthesis tools. Refer to Fig. 11.4 for the synthesis flow with different

Fig. 11.4 Synthesis script processes and indicative commands

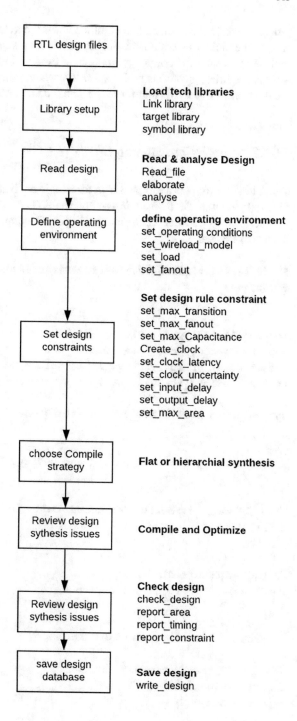

process segments and indicative commands for the synthesis. The designer has to refer to the tool based commands for the processes given in the script segments and replace them with the correct commands to execute in the tool environment. This activity requires a synthesis tool licence to execute. Though the commands resemble the syntax, one needs to refer to the actual commands from the user guide.

11.7.2   Logic Equivalence Check (LEC)

The following script is a sample script for a logic equivalence script. It uses the synthesized netlist as the revised design and the RTL design as the golden reference. The script uses cadence conformal tool-specific commands. This requires a tool license to execute.

```
###############################################################
##########

set log file lec_phy_1801.log -replace
set lowpower option -power_domain_check
set lowpower option -native_1801 -golden_analysis_style post_syn-
thesis -revised_analysis_style post_route

set undefined cell black_box -noascend -both


//Read Library for both Golden and Revised Designs


read library -liberty {standard cell library eg. librarypath/
lib/*}-both


//Read synthesized netlist
read design -verilog -golden top.v

//Read RTL model
read design -verilog -revised top.vg

substitute blackbox model -golden
report design data
report black box

set analyze option -auto
```

```
set system mode lec


// report mapped points
report unmapped points -summary
report unmapped points -extra -unreachable -notmapped

//analyze setup -verbose -effort ultra
add compared points -all

// compare mapped points
compare

// report compare data
report compare data -class nonequivalent -class abort -class
notcompared
report statistics

//********************************************************************
**********
//* Generates the compare data reports
//********************************************************************
**********
tclmode
rm -rf reports
mkdir reports
vpxmode
report compare data -noneq > reports/noneq.rpt
report compare data -abort > reports/abort.rpt
```

Layout Extract File (LEF)

The extract of a LEF is given here. This is a dummy file to show the content of the LEF. This contains the size and electrical parameters of the layer in VLSI. The parasitic extractor from the P&R tool uses this file to extract actual parasitics of the interconnects in the SoC layout for timing and other electrical rule checks (ERCs) during physical design verification.

Extract of the LEF for a particular technology library is shown below:

```
################################################################
#############
LAYER Metal1
 TYPE ROUTING ;
 DIRECTION HORIZONTAL ;
 PITCH 0.19 0.19 ;
```

```
WIDTH 0.06 ;
AREA 0.02 ;
SPACINGTABLE
PARALLELRUNLENGTH 0 0.32 0.75 1.5 2.5 3.5
WIDTH 0 0.06 0.06 0.06 0.06 0.06 0.06
WIDTH 0.1 0.06 0.1 0.1 0.1 0.1 0.1
WIDTH 0.75 0.06 0.1 0.25 0.25 0.25 0.25
WIDTH 1.5 0.06 0.1 0.25 0.45 0.45 0.45
WIDTH 2.5 0.06 0.1 0.25 0.45 0.75 0.75
WIDTH 3.5 0.06 0.1 0.25 0.45 0.75 1.25 ;
MINIMUMCUT 1 WIDTH 0.07 WITHIN 0.3 FROMABOVE ;
MINIMUMCUT 2 WIDTH 0.4 WITHIN 0.3 FROMABOVE ;
MINIMUMCUT 4 WIDTH 1 WITHIN 0.3 FROMABOVE ;
MINIMUMCUT 2 WIDTH 1.5 FROMABOVE LENGTH 1.5 WITHIN 3 ;
MINENCLOSEDAREA 0.045 ;
DIAGSPACING 0.08 ;
DIAGMINEDGELENGTH 0.1 ;
RESISTANCE RPERSQ 0.0736 ;
CAPACITANCE CPERSQDIST 0.0002 ;
THICKNESS 0.15 ;
EDGECAPACITANCE 0.0002 ;
MINIMUMDENSITY 20 ;
MAXIMUMDENSITY 65 ;
DENSITYCHECKWINDOW 120 120 ;
DENSITYCHECKSTEP 60 ;
ANTENNAMODEL OXIDE1 ;
ANTENNAAREARATIO 475 ;
ANTENNACUMAREARATIO 5000 ;
ANTENNACUMDIFFAREARATIO PWL ( ( 0 5000 ) ( 0.099 5000 ) ( 0.1
48045 ) ( 1 48450 ) ) ;
DCCURRENTDENSITY AVERAGE 2 ;
PROPERTY LEF58_SPACING "SPACING 0.08 ENDOFLINE 0.09 WITHIN 0.025
MINLENGTH 0.06 PARALLELEDGE 0.08 WITHIN 0.1 ;" ;
END Metal1

LAYER Via1
 TYPE CUT ;
 SPACING 0.07 ;
 SPACING 0.1 ADJACENTCUTS 3 WITHIN 0.11 ;
 WIDTH 0.07 ;
 ENCLOSURE BELOW 0.005 0.03 ;
 ENCLOSURE ABOVE 0.005 0.03 ;
 ANTENNAMODEL OXIDE1 ;
 ANTENNAAREARATIO 25 ;
 ANTENNADIFFAREARATIO PWL ( ( 0 20 ) ( 1 20 ) ) ;
```

ANTENNACUMROUTINGPLUSCUT ;
ANTENNACUMAREARATIO 180 ;
DCCURRENTDENSITY AVERAGE 0.1 ;
END Via1

11.8 Part III

This part deals with two of the design cases of a block of SoC design for reference. The first one is the mixed block interface (MBI) controller for analog design with PLL and data converters, whose pin diagram is shown in Fig. 11.15.

The internal block diagram is shown in Fig. 11.6.

The second design case is the MINI-SoC for IOT applications and the formal process with documentation for overview, application scenario and, design details of MINI-SoC for IOT are detailed in the following section.

Overview and Application Scenario MINI-SoC can be used for a wide variety of IOT applications like body temperature monitoring device in healthcare, soil humidity monitoring in agriculture, or vehicle tracking device in automobiles by interfacing it to suitable sensor modules and input-output (IO) modules (Fig. 11.7).

MINI-SoC functional requirements

The following are the specifications and requirement for MINI-SoC design.

Intel 8051 processor core with:

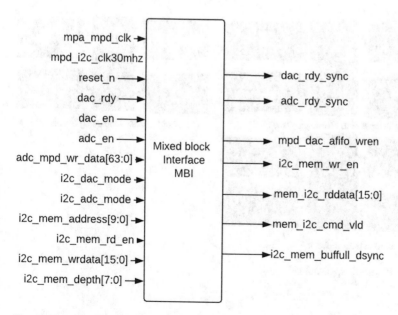

Fig. 11.15 Mixed block interface (MBI)

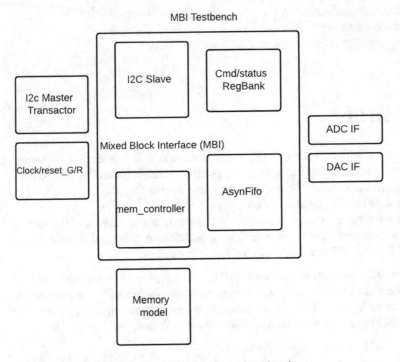

Fig. 11.6 Internal block diagram of MBI design and test bench

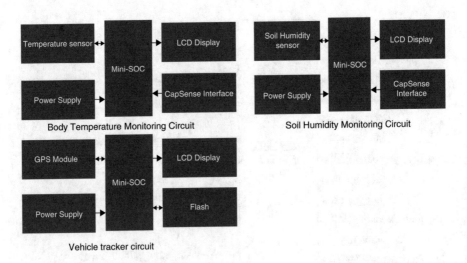

Fig. 11.7 MINI-SoC applications

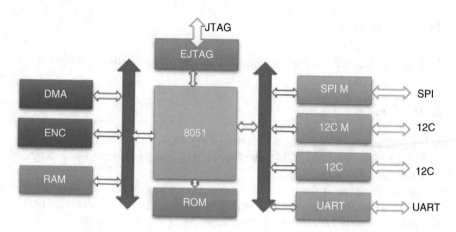

Fig. 11.8 MINI-SoC internal block diagram

1. Power-on reset and programmable brown-out Detection
2. Internal calibrated oscillator
3. External and internal interrupt sources
4. Six sleep modes: idle, ADC noise reduction, power-save, power-down, standby, and extended standby
5. 32K program memory
6. 32K data memory
7. 32 x 8 general-purpose scratchpad registers
8. Master/slave SPI serial interface
9. Byte-oriented 2-wire serial interface (Philips I2 C compatible)
10. Programmable serial UART (Fig. 11.8)

MINI-SoC performance requirements: MINI-SoC should have the following performance requirements:

1. Maximum clock speed of 20 MHz
2. In-System Programming by On-chip Boot Program
3. Powerful Instructions—Most Single Clock Cycle Execution
4. Up to 20 MIPS throughput at 20 MHz

IOs and packaging requirements: {Sample requirement applicable when the design is taken for fabrication}
   28-pin PDIP, 32-lead TQFP
Operating voltage: {Decides library choice when design is taken up for fabrication}

1. 1.8–5.5V
2. Temperature range: -40°C to 85°C
3. Speed grade: –0–20 MHz @ 1.8–5.5V

11.8.1   MINI-SoC Design

This section details the design or microarchitecture of the MINI-SoC design

IO Diagram

MINI-SoC input-output diagram is shown in Fig. 11.9.

MINI-SoC internal block diagram: Fig. 11.10 shows the internal block diagram of MINI-SoC (Table 11.1).

User may register at the weblink **www.opencores.org** and download the MINI-SoC design database from the link https://opencores.org/download/oms8051mini. The SoC subsystem is designed by Mr. Dinesh Annayya, my colleague.

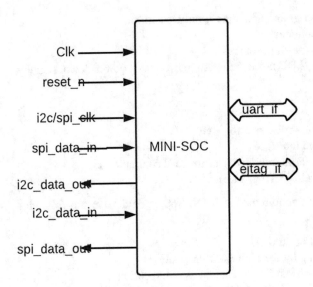

Fig. 11.9 MINI-SoC IO diagram

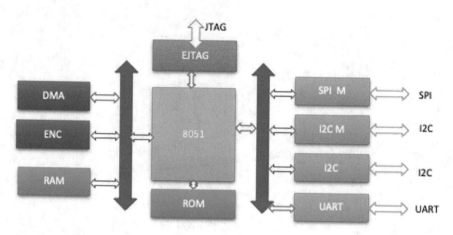

Fig. 11.10 MINI-SoC internal block diagram

Table 11.1 Top-level input-output signals of MINI-SoC

Sl. no	Signal	Width	Direction	Description	Reset value
System interface					
1	clk	1	Input	Clk is the main SoC clock	-
2	reset_n	1	Input	Reset is active low reset signal with which all the internal logic get reset	1'b1
I2c slave interface					
3	I2c_data	1	Inout	I2c data input-output in slave mode	1'b0
4	I2c_clk	1	Input	I2c serial clock input to which i2c data is synchronized in slave mode	1'b0
5	I2c/spi_clk	1	Input	I2c or spi clock input in slave mode which is input by external i2c master	1'b0
6	I2c_sdata	1	Inout	Multiplexed I2c serial data which is in slave mode	1'b0
7	I2c_mdata	1	Inout	Multiplexed I2c serial data which is in master mode	1'b0
8	I2c/spi_mclk	1	Output	I2c or spi clock output in master mode which is generally lower than system clock	1'b0
EJTAG interface					
9	TDI	1	Input	TDI signal	1'b0
10	TDO	1	Output	TDO signal	1'b0
11	TCK	1	Input	Serial JTAG clock	
12	TRST	1	Input	Reset	1'b0
13	TMS	1	Output	Model select	1'b0

Index

© The Editor(s) (if applicable) and The Author(s), under exclusive license to
Springer Nature Switzerland AG 2022
V. S. Chakravarthi, *A Practical Approach to VLSI System on Chip (SOC) Design*,
https://doi.org/10.1007/978-3-031-18363-8

Printed in the United States
by Baker & Taylor Publisher Services